HIP-HOP MATH

CONQUERING WORD PROBLEMS

Alex Kajitani and Alan Sitomer

Scenario-Based Mathematics x Hip-Hop Pedagogy + Student Engagement

Including Standards-Based Strategies to ...

Build Academic Vocabulary
Make a Table
Draw a Picture
Guess and Check
Write an Equation
Work Backwards
Look for Patterns
Make an Organized List

Delivering the 5th Element of Hip-Hop to Capable Young Minds

1. B-boying 2. MC-ing 3. Graffiti Art 4. DJ-ing 5. **Knowledge**

**Note:* All work contained herein begins with the smart and effective implementation of standards-based lesson plans that have been constructed based on three principles:

1. no profanity 2. no homophobia 3. no misogyny

Great vocabulary words to discuss; inappropriate elements to validate in a classroom environment.

Hip-Hop Math: Conquering Word Problems
by Alex Kajitani and Alan Sitomer

Street Smart Press produces innovative, dynamic teaching materials that authentically engage students while simultaneously equipping educators to meet rigorous, high-level academic objectives.

For a host of free, fantastic resources please visit

StreetSmartPress.com
The home of hip-hop pedagogy.

ISBN 978-0-9861546-1-4

First Edition

Cover and Interior Design: Rico Frederick/Jason Lalor
Street Smart Press
3460 Marron Road, Suite 103-160
Oceanside, CA 92056
For all inquiries please visit *www.StreetSmartPress.com*

Typesetters: Wordzworth.com

Contents

Pertinent Research Informing Hip-Hop Pedagogy

* ASCD says that effective reading and writing instruction is anchored in providing the opportunity for students to interpret and compose text about subject areas that they find to be personally meaningful.

* NCTE declares that effective teachers understand the critical importance of adolescents' finding enjoyable texts with which to work and that instruction should center around on learners' authentic real-world interests.

* The U.S. Department of Education explicitly advocates that teachers help students build confidence in their ability to comprehend texts by making literacy experiences more relevant to students' interests and everyday life.

* IRA says that effective reading instruction remains at the core of any successful attempt at educational reform.

* The O.R.C. that says educators who teach reading and writing skills without addressing student engagement are unlikely to yield substantial improvements. The student who refuses to learn will succeed in that effort. On the other hand, students who are motivated to learn can succeed even in less-than-optimal environments.

* The National Center for Education Statistics says studies show that academic achievement is directly associated with engagement in reading and classroom-related activities and that this association is found to hold true for all racial, ethnic, and socioeconomic groups as well as for both gender groups.

* The Department of Justice has documented that the link between academic failure and delinquency, violence, and crime is connected to reading failure and that 85% of juveniles who come in contact with the juvenile court system are functionally illiterate (and more than 70% of inmates in US prisons cannot read above a fourth-grade level).

* The Hip-Hop Center for Research and Teaching at New York University estimates that over three hundred classes on hip-hop are currently being offered at colleges and universities throughout the United States and abroad.

* The Latino Alliance for Literacy Advancement asserts that low literacy skills make people more likely to be poor, sick, drug addicted, alcoholic, criminal, and homeless. As the organization's executive director puts it, "You name the social ill, literacy is the cure."

For more resources and tools, please visit:

STREET SMART PRESS

StreetSmartPress.com
The home of hip-hop pedagogy

Introduction: Why Hip-Hop Math?

by California Teacher of the Year Award Winner Alex Kajitani

It was my first year of teaching, and I was sinking. All that preparation, all those diplomas, and I could not get my middle school students to sit down and pay attention. I felt disrespected and frustrated that I couldn't get them to remember what I had just taught them the day before, yet they could easily remember every word of the new rap song on the radio. Of course, the other problem with this was that they would come in each day singing about violence, drug use, and mistreating women, which frustrated me even more.

And then one afternoon it hit me. Instead of turning off their radios, I needed to offer them a different station. I went home and made up a rap song about the math we were learning at the time (adding and subtracting decimals), calling it "The Itty-Bitty Dot." I practiced it all night, peppered it with clever phrases, and rapped it over an authentic hip-hop beat I'd found online. I remembered my own love of rap in its cleaner youth and imagined how impressed my students would be with my "cool" factor for my way with a rhyme. Early the next morning, when my class came in, I performed it for them ...

The results were disastrous. The students laughed hysterically, and I felt anything but cool—more like a complete flop. Now, not only were they not paying attention, but they were laughing at me. Later that day, I trudged off to lunch like a loser from *The Gong Show*. And then it happened. As I walked by the lunch tables, the students were singing my song! The next day, they eagerly ran into my classroom, saying things like, "Mr. Kajitani, are you going to rap again? Yesterday was the *best day ever* in math class! Are you going to be on MTV?"

From that moment on, everything shifted. I had connected with my students on *their* level, using language they understood to get across what I was trying to teach. I got them interested—even if they were laughing at me, their laughter meant they were present and comfortable (no small feat in the often dangerous neighborhood my students live in). By shifting my approach, I got them excited to come to school, to learn, and to have me as their teacher. Their behavior improved dramatically, and their test scores began to match, and then outpace, their more affluent counterparts.

Of course, this doesn't just happen because you start rapping. But it does happen when you make a commitment, a true commitment, to doing whatever it takes to **stop trying to force our curriculum into the students' lives and, instead, fit our students' lives into the curriculum.**

Our students already love hip-hop. They are deeply influenced by hip-hop's personalities, music, art, clothing, and lifestyle. And at the same time, they don't necessarily love math. In fact, to many students, especially the ones I teach, who come from poverty-stricken neighborhoods where many in their family never graduated from high school, math is a subject many students seek to avoid, never grasping how truly important it is for achieving success in life.

Hip-Hop Math is the bridge that is so desperately needed in mathematics education today. It starts with real-world scenarios and elements of hip-hop that already fascinate today's youth, and then integrates the math in a way that is culturally relevant, pedagogically sound, and truly engaging. None of the math in this book has been "dumbed down"; rather, it has been chunked in a way that teachers can deliver, students can learn, and schools can use, transforming what is possible in education today.

Why Conquering Word Problems?

As the 2009 California Teacher of the Year and a Top-Four Finalist for National Teacher of the Year, I've had the pleasure of visiting hundreds of schools and working with thousands of teachers to improve their instruction in mathematics. At the beginning of each session, I ask everyone in the room to share their biggest challenges when teaching mathematics.

The two most frequent answers I get are these:

1. Students' lack of knowledge of basic math facts
2. Students' inability to read and persevere in solving word problems

For years, we've been teaching word problems as something we do after drilling students on procedures. In a typical homework assignment, we assign #1–25, with numbers 23, 24, and 25 being word problems, usually carrying equal weight to the simpler procedures at the beginning of the section.

Students who are already reluctant readers, or who are English language learners, often struggle the most with word problems. The same goes for students who fail to see the relevance to their own lives of the situations presented. Does a kid growing up in the urban setting of the Bronx really care about the hypothetical farmer who needs to decide on the length and width of his fence to maximize the total grazing area for his cows?

Conquering Word Problems is about taking the hip-hop students love and combining it with the math they need to know. It is a book that leverages what excites students about life to transform them into confident, articulate scholars who think critically and solve problems in the twenty-first century.

Best of luck with all your efforts.

Alex Kajitani
COO, Street Smart Press

How a Language Arts Person Like Myself Came to Coauthor a Math Book

by California Teacher of the Year Award Winner Alan Sitomer

What happens when a literacy specialist who has fashioned a career from elevating the academic performance of capable but disengaged students sits down with a like-minded, awesomely talented math teacher and has a cup of coffee?

This book.

As an "English" guy, I have long stared in amazement at the math assessments taking place across the country. "These are reading comprehension problems before they are evaluations of mathematical abilities," I'd declare to anyone who'd listen.

Many nodded, yet, after more than a decade of head tilts, few things changed.

To a person in the world of ELA, the obvious amount of pure literacy skills necessary to tackle word problems in math class seemed self-evident. But on most campuses, as we all know, literacy exists in one silo of school and math exists in another … and rarely do the two ever break bread.

Alex knew, and I knew, that if we were ever going to improve our kids' competency and performance when it came to word problems, a math teacher and an ELA teacher were going to need to sit down and tear down the artificial walls that separated our two disciplines.

We have now done that, but let me tell you, writing this book was hard. Much harder than I imagined, in fact. Not because the math proved beyond my abilities; I'm no Pythagoras, but I can pull off a little bit of trig. No, this book required a Herculean amount of effort from my side of the fence because writing a cogent, crisp, well-executed mathematical word problem requires an incredible amount of precision.

In fact, one thing became apparent very early on in the drafting of this text: many kids are underperforming in math because so, so many of the word problems they are being asked to answer are just *horribly* written. Questions are sloppy, ambiguous, imprecise, and lazy. Researching this text proved absolutely maddening for me. I was, and still am, appalled at the work of other publishers.

"No wonder so many kids are faring so poorly! Look at how convoluted this question is," I'd yell. "These problems STINK!"

Alex, after reducing my coffee intake and walking me back off the ledge, would remind me that this was why we were taking such great pains to write a great text. Excellence would be our answer.

I am supremely proud of this book. And I deeply believe that it might prove to be a key that unlocks both comprehension as well as enjoyment of math for many, many of your students.

Godspeed and good luck to you. Many of us are rooting for your success.

Alan Sitomer

CEO, Street Smart Press

THE SIX-SQUARE STRATEGY TO CONQUERING WORD PROBLEMS

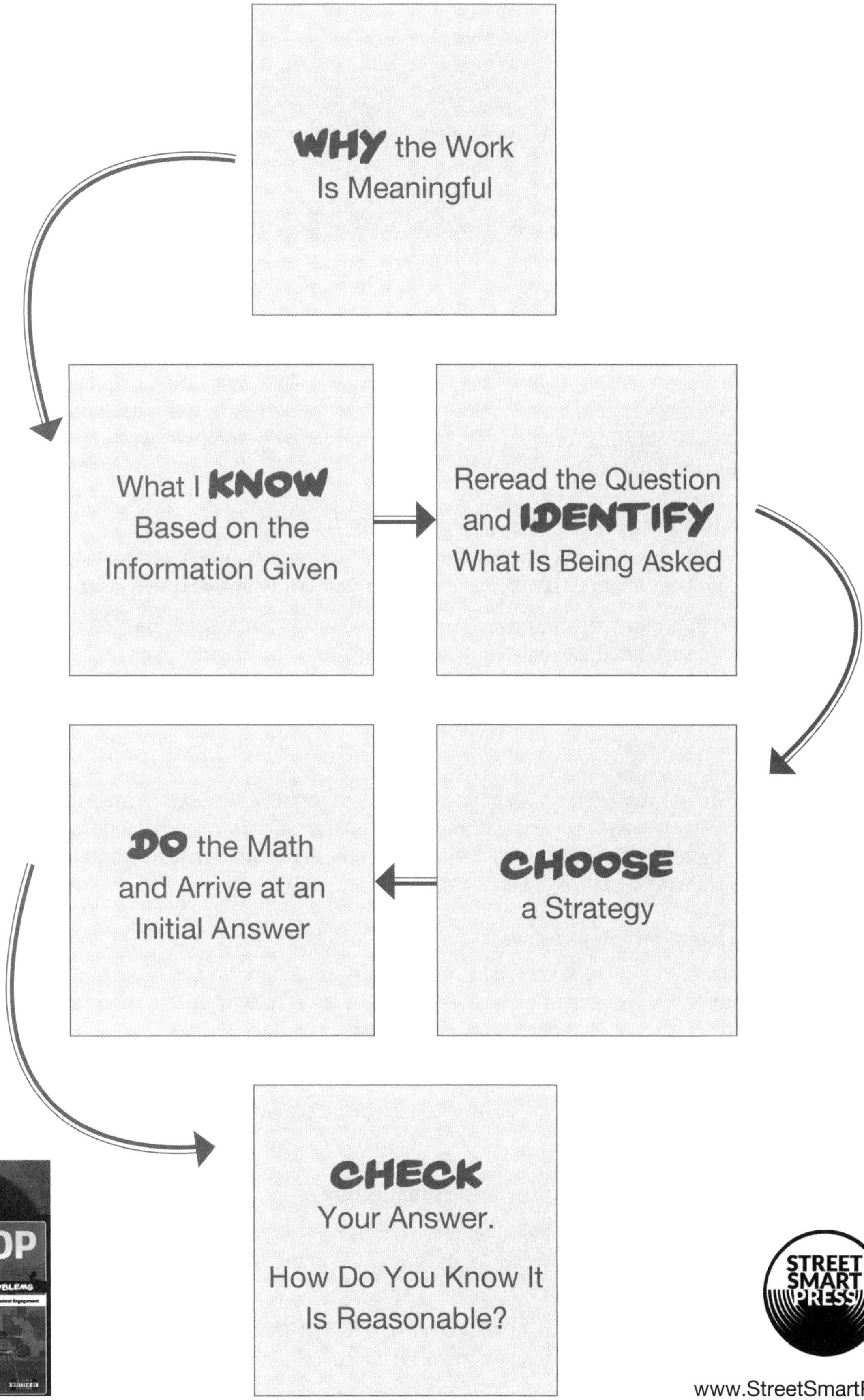

www.StreetSmartPress.com

THE SIX-SQUARE STRATEGY TO CONQUERING WORD PROBLEMS EXPLAINED

WHY the Work Is Meaningful

"Why do I even need to know this?" is a commonly asked question by students in math class. Have an answer. One that is overt, transparent, and true. Math is important to every last one of us, and when students see the critical role that math plays in the career success of their favorite hip-hop artists, a real-world connection is made between classroom activities and life.

Research clearly shows that when students are internally motivated to pursue meaningful, purposeful activities, achievement soars. This is why Hip-Hop Math taps into the power of hip-hop pedagogy to provide culturally relevant, scenario-based, student-centered problem-solving activities. We know that when kids find meaning in the work, they do a better job at the work.

Remember, **before students can solve for "X," they need to know the "*Why!*"**

What I KNOW Based on the Information Given

The first step to effective problem solving is assessing what is known. This is why the second stage of the Six-Square strategy has students comb through the scenario they have been given to **determine the actual information** being provided. Identifying facts, avoiding unfounded assumptions, and taking good stock of all data, information, and statistics are critical to unpacking word problems effectively.

Reread the Question and IDENTIFY What Is Being Asked

All word problems have one thing in common: there is something that needs to be solved. This is why the third step of the Six-Square Strategy places direct emphasis on asking students to **clearly identify what they are being asked to answer**.

Many, many students struggle to solve word problems because they do not answer the question that is being asked of them (and instead respond to the question they assume they are being asked, which is often an entirely different thing). This is where literacy and reading comprehension come into play. To be successful, students must know exactly what they are being asked.

CHOOSE a Strategy

How a student chooses to solve a word problem is of utmost importance. Struggling students will often rush into completing an operation without first determining whether that operation will get them the desired result. In the fourth step of the Six-Square Strategy, **students survey a wide range of available strategies and then choose the right one(s)** that will help them effectively arrive at the correct solution.

DO the Math and Arrive at an Initial Answer

Mathematical word problems are solved by translating language into numbers or figures and then "doing the math" (i.e., performing the calculations). In the fifth step of the Six-Square Strategy, **students use their procedural knowledge** to complete the operations necessary to arrive at an initial answer.

Not their final answer—more on that in a moment—but their initial response.

Note: In this step, effective students will make an effort to produce work that is neat and organized to reduce unnecessary mistakes.

CHECK Your Answer. How Do You Know It Is Reasonable?

Simply telling students "Check your work" is only effective if they have done their math correctly. By asking students to explain how they know their answer is reasonable, the sixth step of the Six-Square Strategy ensures that **students prove why their solutions are reasonable**.

An added benefit of this stage is that asking students to explain the reasonability of their answers deepens their understanding of the problem-solving process as a whole (often by comparing their answer to other possible answers). Constantly asking themselves "Does this make sense?" is a critical component of success with scenario-based mathematics.

UNIT 1

MAKE A TABLE

Making a table helps you:

organize information ▪ identify patterns ▪ see relationships

Step 1: What are you comparing?

These are your labels.
The number of labels is the number of columns you'll need.

Step 2: What goes in the first row (this is the info that is given)?

Step 3: How many rows will you need?

(Hint: While the problem will often tell you how many columns you'll need, the number of rows you'll need will often NOT be provided. Thus, the mnemonic below: "When in doubt, 2 x 7 is the route.)

Step 4: Build the table:

When in doubt, 2 x 7 is the route! (That means 2 columns and 7 rows.)

Label 1	Label 2

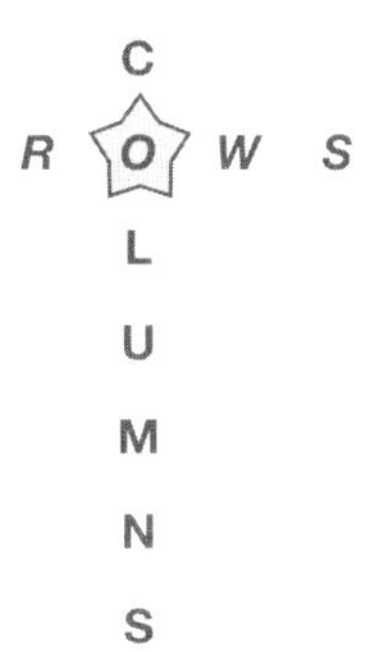

Step 5: Continue filling in until you find the desired result.

Step 6: How do you know your answer is reasonable?

HIP-HOP MATH STUDENT CHALLENGE

Recording an Album with Eminem, Week 1

Scenario: Eminem is recording a new album. The producer needs to estimate a budget. He approximates that it will take Eminem 9 hours of studio time to record 5 songs. At this rate, how many songs could he record in 27 hours?

Step 1: What are you comparing?

Step 2: What goes in the first row (this is the info that is given)?

Step 3: How many rows will you need?

Step 4: Build the table:

# Hours	# Songs

Step 5: Continue filling in until you find the desired result.

Step 6: How do you know your answer is reasonable?

HIP-HOP MATH ANSWER GUIDE

Recording an Album with Eminem, Week 1

Scenario: Eminem is recording a new album. The producer needs to estimate a budget. He approximates that it will take Eminem 9 hours of studio time to record 5 songs. At this rate, how many songs could he record in 27 hours?

Step 1: What are you comparing? *# of hours and # of songs (These are my labels.)*

Step 2: What goes in the first row (this is the info that is given)? *9 (under hours) and 5 (under songs)*

Step 3: How many rows will you need? *Not sure, so I'll start with 7. (When in doubt, 2 x 7 is the route!)*

Step 4: Build the table:

# Hours	# Songs
9	5

But 9 is not enough, so I'll do 9 more hours which should result in 5 more songs.

Step 5: Continue filling in until you find the desired result.

# Hours	# Songs
9	5
18	10

18 hours isn't the number I'm looking for, so I'll do 9 more hours...

# Hours	# Songs
9	5
18	10
27	15

There it is! It will take 27 hours to record 15 songs.

Step 6: How do you know your answer is reasonable?

(Answers will vary.) For hours, 9 goes into 27 three times; and for songs, 5 goes into 15 three times.

HIP-HOP MATH STUDENT CHALLENGE

Recording an Album with Eminem, Week 2

Scenario: Eminem has now decided to turn this into a double album. On Monday morning of week 2, his producer estimates that it will take Eminem 7 hours to record 4 more songs. At this rate, how many songs can Eminem record in 35 hours of studio time?

Step 1: What are you comparing?

Step 2: What goes in the first row (this is the info that is given)?

Step 3: How many rows will you need?

Step 4: Build the table:

# Hours	# Songs

Step 5: Continue filling in until you find the desired result.

Step 6: How do you know your answer is reasonable?

___

___

___

HIP-HOP MATH ANSWER GUIDE

Recording an Album with Eminem, Week 2

Scenario: Eminem has now decided to turn this into a double album. On Monday morning of week 2, his producer estimates that it will take Eminem 7 hours to record 4 more songs. At this rate, how many songs can Eminem record in 35 hours of studio time?

Step 1: What are you comparing? *# of hours and # of songs (my labels)*

Step 2: What goes in the first row (this is the info that is given)? *7 (under hours) and 4 (under songs)*

Step 3: How many rows will you need? *Not sure, so I'll start with 2 x 7.*

Step 4: Build the table:

# Hours	# Songs
7	4

Step 5: Continue filling in until you find the desired result.

# Hours	# Songs
7	4
14	8
21	12
28	16
35	20

But 7 hours is not enough, so I'll do 7 more, resulting in 4 more songs.

Still not enough, I'll do 7 more hours & 4 more songs.

Getting there, need 7 more hours.

Almost there...

There it is! It will take 35 hours to record 20 songs.

Step 6: How do you know your answer is reasonable?

(Answers will vary.) For hours, 7 goes into 35 five times; and for songs, 4 goes into 20 five times.

HIP-HOP MATH STUDENT CHALLENGE

Recording an Album with Eminem, Week 3

Scenario: Eminem also decides that he wants to bring in a female R&B vocalist to sing on 6 of his new songs. She is able to record 2 songs in 1.5 hours. At this rate, how many hours will it take her to record over all 6 of the songs?

Step 1: What are you comparing?

Step 2: What goes in the first row (this is the info that is given)?

Step 3: How many rows will you need?

Step 4: Build the table:

Step 5: Continue filling in until you find the desired result.

Step 6: How do you know your answer is reasonable?

HIP-HOP MATH ANSWER GUIDE

Recording an Album with Eminem, Week 3

Scenario: Eminem also decides that he wants to bring in a female R&B vocalist to sing on 6 of his new songs. She is able to record 2 songs in 1.5 hours. At this rate, how many hours will it take her to record over all 6 of the songs?

Step 1: What are you comparing? *# of songs and # of hours (my labels)*

Step 2: What goes in the first row (this is the info that is given)? *2 (under hours) and 1.5 (under songs)*

Step 3: How many rows will you need? *Not sure, so I'll start with 2 x 7.*

Step 4: Build the table:

# Songs	# Hours
2	1.5

Step 5: Continue filling in until you find the desired result.

# Songs	# Hours
2	1.5
4	3.0
6	4.5

But 2 songs is not enough, so I'll do 2 more resulting in 1.5 more hours.

Almost there. I need to get to 6 songs so I'll do 2 more.

So recording over 6 songs will take 4.5 hours. That's 4 hours and 30 minutes.

Step 6: How do you know your answer is reasonable?

(Answers will vary.) For hours, 2 goes into 6 three times; and for songs, 1.5 goes into 4.5 three times.

HIP-HOP MATH STUDENT CHALLENGE

The Album Production Team, Week 4

Scenario: Even though Eminem recorded 35 total songs, only 24 of the tracks will be used on his new release. The album production team will now spend the new few days mixing and mastering the material. Eminem's chief sound engineer makes $250 per hour. It takes her 6 hours to master 4 songs. At this rate, how much will she be paid to master all 24 songs?

How do you know your answer is reasonable?

____

____

____

HIP-HOP MATH ANSWER GUIDE

The Album Production Team, Week 4

Scenario: Even though Eminem recorded 35 total songs, only 24 of the tracks will be used on his new release. The album production team will now spend the new few days mixing and mastering the material. Eminem's chief sound engineer makes $250 per hour. It takes her 6 hours to master 4 songs. At this rate, how much will she be paid to master all 24 songs?

Step 1: What are you comparing? *# of hours and # of songs (my labels)*

Step 2: What goes in the first row (this is the info that is given)? *6 (under hours) and 4 (under songs)*

Step 3: How many rows will you need? *Not sure, so I'll start with 2 x 7.*

Step 4: Build the table:

# Hours	# Songs
6	4

Step 5: Continue filling in until you find the desired result.

# Hours	# Songs
6	4
12	8
18	12
24	16
30	20
36	24

I need to get to 24 songs, so I'll keep adding 6 hours, and 4 songs until I get there.

I'll keep going...

Almost there...

So it takes the sound engineer 36 hours to master all 24 songs.

I still need to multiply by her hourly rate: $250 x36

She'll be paid $9,000

Step 6: How do you know your answer is reasonable?

*(**Answers will vary.**) 10 hours of work would be $2,500, and 40 hours would be $10,000. 36 hours is in between 10 and 40, just like $9,000 is in between $2,500 and $10,000.*

HIP-HOP MATH STUDENT CHALLENGE

Recording an Album with Eminem, Week 5

Scenario: To get this album done quickly, the sound engineer hires an assistant and pays the assistant $140 per 4 songs. Based on the previous question, how much will the assistant be paid to master all of the songs on the double album?

How do you know your answer is reasonable?

____

____

____

HIP-HOP MATH ANSWER GUIDE

Recording an Album with Eminem, Week 5

Scenario: To get this album done quickly, the sound engineer hires an assistant and pays the assistant $140 per 4 songs. Based on the previous question, how much will the assistant be paid to master all of the songs on the double album?

Step 1: What are you comparing? *Dollars ($), and # of songs (my labels)*

Step 2: What goes in the first row (this is the info that is given)? *140 (under dollars) and 4 (under songs)*

Step 3: How many rows will you need? *Not sure, so I'll start with 2 x 7.*

Step 4: Build the table:

Dollars	# Songs
$140	4

Step 5: Continue filling in until you find the desired result.

Dollars	# Songs
$140	4
$280	8
$420	12
$560	16
$700	20
$840	24

I need to get to 24 songs, so I'll keep adding $140, and 4 more songs until I get there!

I'm half-way there!

Almost there...

So the assistant will be paid $840 to master all 24 songs on the double album.

Step 6: How do you know your answer is reasonable?

(Answers will vary.) I also figured out that $140 for 4 songs is $35 per song, and $35 x 24 songs is $840.

HIP-HOP MATH STUDENT CHALLENGE

Recording an Album with Eminem, Week 6

Scenario: Eminem's producer must hire a graphic designer to complete the album cover.

After a long search, it's now down to two candidates...

The first candidate, Jonathan Mannion, an art director who has worked with Jay-Z and DMX, charges $200 per album plus $80 per hour. The second graphic artist, Kimberly Holt, an art director who has worked with Dr. Dre and Snoop Dogg, charges $260 per album plus $65 per hour.

QUESTION A: Who would be less expensive to hire if it takes 6 hours to design Eminem's new album cover?

QUESTION B: At what number of hours do the two graphic designers cost the same amount of money?

HIP-HOP MATH ANSWER GUIDE

Recording an Album with Eminem, Week 6

Scenario: Eminem's producer must hire a graphic designer to complete the album cover.

After a long search, it's now down to two candidates...

The first candidate, Jonathan Mannion, an art director who has worked with Jay-Z and DMX, charges $200 per album plus $80 per hour. The second graphic artist, Kimberly Holt, an art director who has worked with Dr. Dre and Snoop Dogg, charges $260 per album plus $65 per hour.

QUESTION A: Who would be less expensive to hire if it takes 6 hours to design Eminem's new album cover?

* Comparing: # of hours, Jonathan Mannion (JM), and Kimberly Holt (KH) (my labels)

* # of Columns: 3 (since there are 3 labels)

* # of Rows: Not sure, so I'll start with 3 x 7 (but I think 6, since it takes 6 hours to design the album cover).

Hours	JM ($ Cost)	KH ($ Cost)
1	200 + 80 = 280	260 + 65 = 325
2	200 + 160 = 360	260 + 130 = 390
3	200 + 240 = 440	260 + 195 = 455
4	200 + 320 = 520	260 + 260 = 520
5	200 + 400 = 600	260 + 325 = 585
6	200 + 480 = 680	260 + 390 = 650

Kimberly Holt is cheaper by $30.

QUESTION B: At which point on an hourly pay scale do the two graphic designers cost the same amount of money?

At 4 hours, Jonathan and Kimberly both cost $520.

REAL-WORLD SCENARIO: "TURN TABLES" INTO WORD PROBLEMS

PLATINUM PROBLEM

Scenario: The record company has a hot new artist it has just signed and a maximum budget of $15,000 to produce an entire album. There are 12 songs on the new album, and it takes an average of 2 hours to record and master each song. Plus, the album needs cover art, and that's going to require 7 hours of a graphic designer's time.

Below is a table that shows the different prices to rent the studio:

Studio Hours	Cost (Bring Your Own Engineer)	Cost with Our Studio's Engineer
1 hour	$80	$230
2 hours	$150	$430
3 hours	$210	$600
½ day (4 hours)	$260	$760
Full day (8 hours)	$500	$1,380
Weekly rate (5 full days)	$2,250	$6,250

Below is a table that shows the prices of your two top picks, if you decide to hire your own sound engineer:

Sound Engineer	Max "the Chef" Walker	Cindy "Ms. Mic" Jenkins
Hourly pay rate	$200	$165
Weekly discount rate (40 hours in a week)	$6,400	none

Below is a table that shows the prices of your two top picks, if you decide to hire your own graphic artist:

Graphic Artist	CJ "Rock" Anderson	Regina "the Ink" Flores
One-time album setup fee	$260	$200
Hourly pay rate	$65	$80

HIP-HOP MATH STUDENT CHALLENGE

Real-World Scenario: "Turn Tables" into Word Problems

Use the tables provided to write 3 original word problems.
(Then give them to a friend to try to solve. NOTE: You must be able to solve your own problems!)

Problem #1: ______________________________

Problem #2: ______________________________

Problem #3: ______________________________

HIP-HOP MATH ANSWER GUIDE

Real-World Scenario: "Turn Tables" into Word Problems

PLATINUM PROBLEM

Use the tables provided to write 3 original word problems.
(Then give them to a friend to try to solve. NOTE: You must be able to solve your own problems!)

Problem #1

(Answers will vary.) If it takes 2 hours to record and master each song, how much will it cost to do this for all songs if you decide to use the studio's engineer?

Problem #2

(Answers will vary.) If you decide to hire Max Walker as your sound engineer, is it better to pay him his hourly rate or his weekly rate if you only plan on using him for 33 hours that week?

Problem #3

(Answers will vary.) How much (if any) money will you have left in your budget if you rent the studio for the entire week and hire Cindy Jenkins as your sound engineer and CJ Anderson as your graphic artist?

TWELVE TOP TERMS: KEY ACADEMIC VOCABULARY

Budget
An estimate of income and expenditure for a set period of time.

Column
A vertical division of a table, page, or text.

Compare
To estimate, measure, or note the similarity or dissimilarity between.

Estimate
An approximate calculation or judgment of the value, number, or quantity of something.

Organize
To arrange into a certain or structured way.

Pattern
A sequence of numbers that can be used.

Rate
A measure or quantity, typically measured against another measure or quantity.

Ratio
A relationship between two numbers. It tells you how much of one thing there is compared to another thing.

Reasonable
A number that makes sense in the problem or scenario.

Relationship
The way in which two or more numbers, concepts, objects, or people are connected.

Row
A horizontal division of a table, page, or text.

Table
A set of facts or figures systematically displayed, especially in columns and rows.

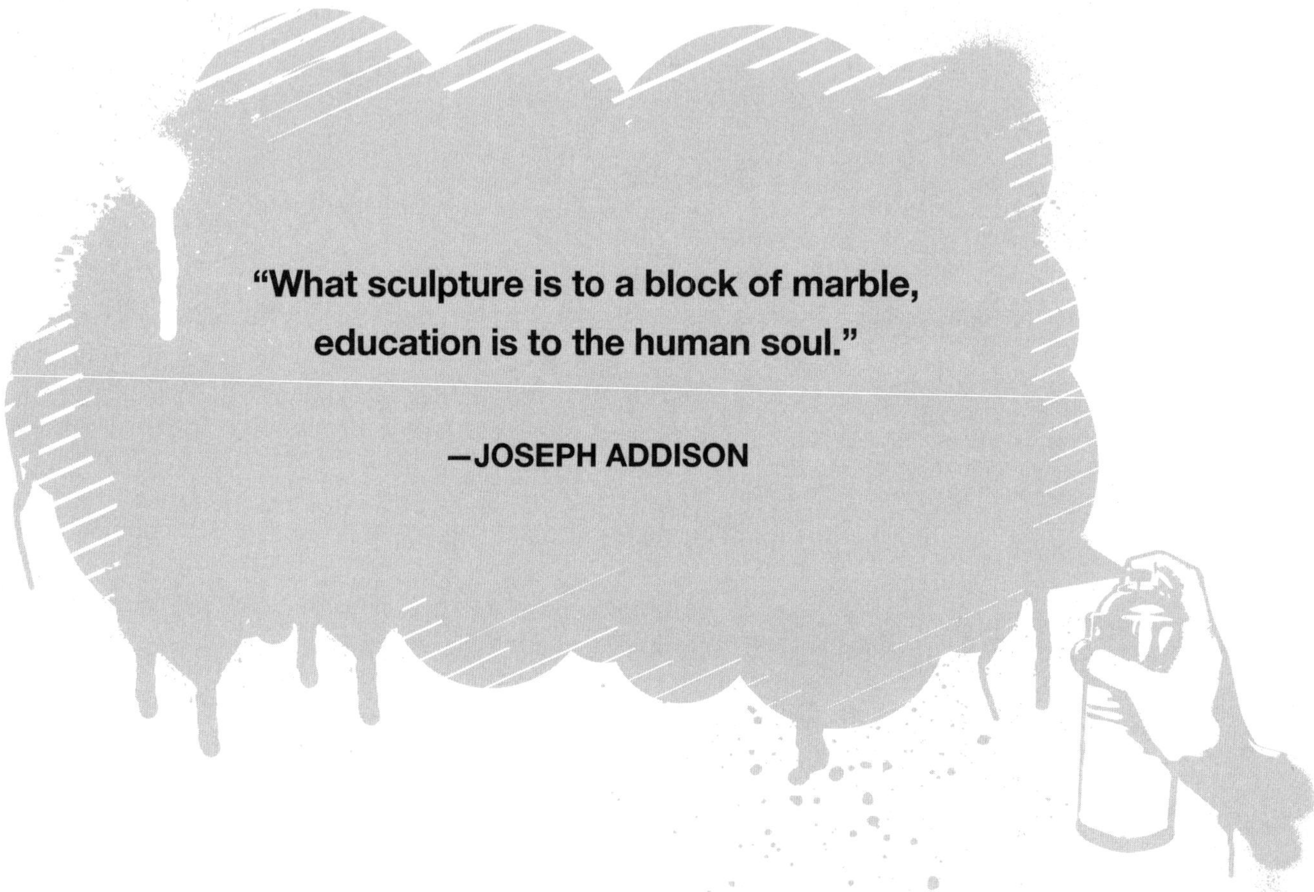

"What sculpture is to a block of marble, education is to the human soul."

—JOSEPH ADDISON

UNIT 2

WRITE AN EQUATION

Writing an equation helps you:

identify what you're looking for ▪ solve a problem by changing words into numbers and math symbols ▪ decide which operation to use

Step 1: What are you trying to find?

Step 2: Assign a variable (such as *x*) to that quantity.

Step 3: Translate words into numbers, symbols, and operations. This will build your equation!

Step 4: Solve the equation.

Step 5: Replace the variable with your answer and check that it makes the equation true.

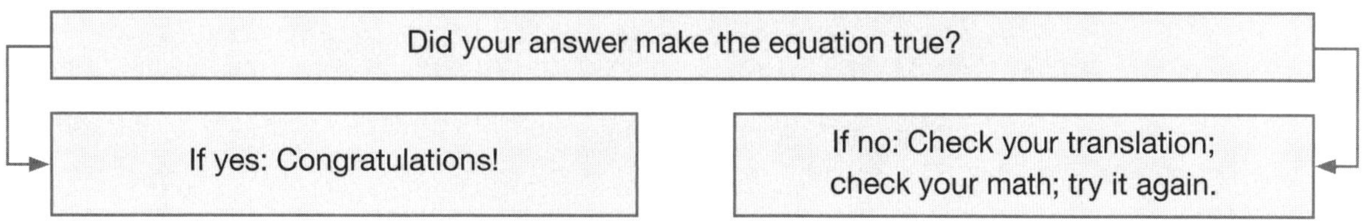

Step 6: How do you know your answer is reasonable?

THE 360 RECORD DEAL

A 360 deal is a business relationship between an artist and a music industry company. The company agrees to provide financial support for the artist, including direct advances and funds for marketing, promotion, and touring. The artist agrees to give the company a percentage of all income, including sales of recorded music, honoraria for live performances, and any other streams of revenue.

In 2008, Jay-Z signed a 360 deal with industry giant Live Nation for $150 million (over 10 years). Live Nation agreed to oversee such areas as touring, concert promotion, merchandising, ticketing, and clothing sales.

HIP-HOP MATH STUDENT CHALLENGE

Breaking Down the Deal

According to a New York Times article, Live Nation is expected to contribute:

- $25 million for overhead costs
- $20 million for publishing, licensing, and other rights
- $50 million for up-front costs (including a general advance)
- $10 million per new album (minimum of 3 albums)
- $25 million to finance Jay-Z's future acquisitions or investments

Scenario: Jay-Z needs to hire an office staff for his new enterprise. If he pays an average salary of $60,000 per person, how many people can he hire with annual budget of $3 million per year for staffing?

Step 1: What are you trying to find?

Step 2: Assign a variable (such as *x*) to that quantity.

Step 3: Translate words into numbers and symbols. This will build your equation!

Step 4: Solve the equation.

Step 5: Replace the variable with your answer and check that it makes the equation true.

Did your answer make the equation true?

If yes: Congratulations!

If no: Check your translation; check your math; try it again.

Step 6: How do you know your answer is reasonable?

HIP-HOP MATH ANSWER GUIDE

Breaking Down the Deal

According to a New York Times article, Live Nation is expected to contribute:

- $25 million for overhead costs
- $20 million for publishing, licensing, and other rights
- $50 million for up-front costs (including a general advance)
- $10 million per new album (minimum of 3 albums)
- $25 million to finance Jay-Z's future acquisitions or investments

Scenario: Jay-Z needs to hire an office staff for his new enterprise. If he pays an average salary of $60,000 per person how many people can he hire with annual budget of $3 million per year for staffing?

Step 1: What are you trying to find? *The number of people Jay-Z can hire.*

Step 2: Assign a variable (such as *x*) to that quantity. x

Step 3: Translate words into numbers and symbols. This will build your equation!

(I'll take the dollar sign out until I solve the equation, then put it back in if I need to in the answer.)

$$60{,}000x = 3{,}000{,}000$$

Step 4: Solve the equation.

→ $\frac{60{,}000x}{60{,}000} = \frac{3{,}000{,}000}{60{,}000}$

→ $x =$ *50 people*

Step 5: Replace the variable with your answer and check that it makes the equation true.

→ $60{,}000\,(50) = 3{,}000{,}000$

→ $3{,}000{,}000 = 3{,}000{,}000$

Did your answer make the equation true? *Yes*

If yes: Congratulations!

If no: *Check your translation; check your math; try it again.

Step 6: How do you know your answer is reasonable?

(Answers will vary.) 100 people would cost $6 million, and since 50 is half of 100, half of $6 million is $3 million.

HIP-HOP MATH STUDENT CHALLENGE

Breaking Down the Deal

According to a New York Times article, Live Nation is expected to contribute:

- $25 million for overhead costs
- $20 million for publishing, licensing, and other rights
- a payment of $50 million for up-front costs (including a general advance)
- a payment of $10 million per new album (minimum of 3 albums)
- $25 million to finance Jay-Z's future acquisitions or investments

Scenario: Jay-Z also needs to hire a road crew for his upcoming tour. If he pays an average salary of $50,000 per person, how many people can he hire with annual budget of $2 million per year for staffing?

Step 1: What are you trying to find?

Step 2: Assign a variable (such as *x*) to that quantity.

Step 3: Translate words into numbers and symbols. This will build your equation!

Step 4: Solve the equation.

Step 5: Replace the variable with your answer and check that it makes the equation true.

Did your answer make the equation true?

If yes: Congratulations!

If no: Check your translation; check your math; try it again.

Step 6: How do you know your answer is reasonable?

____

____

HIP-HOP MATH ANSWER GUIDE

Breaking Down the Deal

According to a New York Times article, Live Nation is expected to contribute:

- $25 million for overhead costs
- $20 million for publishing, licensing, and other rights
- a payment of $50 million for up-front costs (including a general advance)
- a payment of $10 million per new album (minimum of 3 albums)
- $25 million to finance Jay-Z's future acquisitions or investments

Scenario: Jay-Z also needs to hire a road crew for his upcoming tour. If he pays an average salary of $50,000 per person, how many people can he hire with annual budget of $2 million per year for staffing?

Step 1: What are you trying to find? *The number of people Jay-Z can hire.*

Step 2: Assign a variable (such as *x*) to that quantity. *x*

Step 3: Translate words into numbers and symbols. This will build your equation!

(I'll take the dollar sign out until I solve the equation, then put it back in if I need to in the answer.)

$$50{,}000x = 2{,}000{,}000$$

Step 4: Solve the equation.

→ $\frac{50{,}000x}{50{,}000} = \frac{2{,}000{,}000}{50{,}000}$

→ $x =$ *40 people*

Step 5: Replace the variable with your answer and check that it makes the equation true.

→ $50{,}000\,(40) = 2{,}000{,}000$

→ $2{,}000{,}000 = 2{,}000{,}000$

Did your answer make the equation true? *Yes*

If yes: Congratulations!

If no: *Check your translation; check your math; try it again.

Step 6: How do you know your answer is reasonable?

(Answers will vary.) Every 20 people will cost $1 million, so 40 people will cost $2 million.

HIP-HOP MATH STUDENT CHALLENGE
Investing in the Deal, Part 1

Scenario: Jay-Z is thinking about using part of the $25 million that he has available in his acquisitions account to invest in 3 different businesses: a sneaker company for $3 million, a technology company for $6 million, and a sports drink company. After these three expenditures, he will still have $12 million left for future opportunities. How much will he be investing in the sports drink company?

Step 1: What are you trying to find?

Step 2: Assign a variable (such as *x*) to that quantity.

Step 3: Translate words into numbers and symbols. This will build your equation!

Step 4: Solve the equation.

Step 5: Replace the variable with your answer and check that it makes the equation true.

Did your answer make the equation true?

If yes: Congratulations!

If no: Check your translation; check your math; try it again.

Step 6: How do you know your answer is reasonable?

HIP-HOP MATH ANSWER GUIDE

Investing in the Deal, Part 1

Scenario: Jay-Z is thinking about using part of the \$25 million that he has available in his acquisitions account to invest in 3 different businesses: a sneaker company for \$3 million, a technology company for \$6 million, and a sports drink company. After these three expenditures, he will still have \$12 million left for future opportunities. How much will he be investing in the sports drink company?

Step 1: What are you trying to find? *The amount of money Jay-Z will invest in the sports drink company.*

Step 2: Assign a variable (such as *x*) to that quantity. *d (for "drink company")*

Step 3: Translate words into numbers and symbols. This will build your equation!

(I'll remove the dollar sign and put the amounts in millions; this will make the amounts easier to work with.)

$$3 + 6 + d + 12 = 25$$

Step 4: Solve the equation.

$3 + 6 + d + 12 = 25$ *(combine like terms)*

→ $21 + d = 25$

$-21 \qquad -21$

→ $d = 4$ *Jay-Z will invest \$4 million in the drink company.*

Step 5: Replace the variable with your answer and check that it makes the equation true.

$3 + 6 + 4 + 12 = 25$

→ $25 = 25$

Did your answer make the equation true? *Yes*

If yes: Congratulations!

If no: Check your translation; check your math; try it again.

Step 6: How do you know your answer is reasonable?

(Answers will vary.) \$4 million is close to the amounts Jay-Z will be investing in the sneaker company (\$3 million) and the tech company (\$6 million).

HIP-HOP MATH STUDENT CHALLENGE

Investing in the Deal, Part 2

Scenario: Jay-Z would like to invest the remaining $12 million in developing up-and-coming hip-hop talent. Investing in a rap group costs $3 million, and investing in a solo artist costs $2 million. If he invests in two rap groups, how many solo artists can he invest in?

Step 1: What are you trying to find?

Step 2: Assign a variable (such as *x*) to that quantity.

Step 3: Translate words into numbers and symbols. This will build your equation!

Step 4: Solve the equation.

Step 5: Replace the variable with your answer and check that it makes the equation true.

Did your answer make the equation true?

If yes: Congratulations!

If no: Check your translation; check your math; try it again.

Step 6: How do you know your answer is reasonable?

HIP-HOP MATH ANSWER GUIDE

Investing in the Deal, Part 2

Scenario: Jay-Z would like to invest the remaining $12 million in developing up-and-coming hip-hop talent. Investing in a rap group costs $3 million, and investing in a solo artist costs $2 million. If he invests in two rap groups, how many solo artists can he invest in?

Step 1: What are you trying to find? *The number of solo artists Jay-Z can invest in.*

Step 2: Assign a variable (such as x) to that quantity. *a (for "artist")*

Step 3: Translate words into numbers and symbols. This will build your equation!

(I'll remove the dollar sign, and put the amounts in millions; this will make the amounts easier to work with.)

$$3(2) + 2(a) = 12$$

Step 4: Solve the equation.

$$3(2) + 2(a) = 12$$

$$\rightarrow \quad 6 + 2a = 12$$

$$-6 \qquad -6$$

$$\rightarrow \quad \frac{\cancel{2}a}{\cancel{2}} = \frac{6}{2}$$

$$\rightarrow \quad a = \text{3 solo artists}$$

Step 5: Replace the variable with your answer and check that it makes the equation true.

$$3(2) + 2(3) = 12$$

$$\rightarrow \quad 6 + 6 = 12$$

$$\rightarrow \quad 12 = 12$$

Did your answer make the equation true? *Yes*

If yes: Congratulations!

If no: Check your translation; check your math; try it again.

Step 6: How do you know your answer is reasonable?

*(**Answers will vary.**) Investing in 3 rap groups would cost $9 million, so investing in 3 solo artists would cost less ($6 million).*

HIP-HOP MATH STUDENT CHALLENGE

Closing the Deal, Part 1

Scenario: To complete the deal, Jay-Z is thinking about hiring a high-powered attorney who will be paid on a sliding scale. Instead of a salary, the lawyer will receive a percentage of Jay-Z's total contract from the $150 million LiveNation 360 deal. The structure of the fees is as follows:

- 5% for the first $50 million
- 4% for the next $40 million
- 3% for all monies thereafter

How much money does the lawyer earn at the 3% commission rate?

How do you know your answer is reasonable?

HIP-HOP MATH ANSWER GUIDE

Closing the Deal, Part 1

Scenario: To complete the deal, Jay-Z is thinking about hiring a high-powered attorney who will be paid on a sliding scale. Instead of a salary, the lawyer will receive a percentage of Jay-Z's total contract from the $150 million LiveNation 360 deal. The structure of the fees is as follows:

- 5% for the first $50 million
- 4% for the next $40 million
- 3% for all monies thereafter

How much money does the lawyer earn at the 3% commission rate?

Step 1: What are you trying to find? *The amount the lawyer earns at a 3% commission rate.*

Step 2: Assign a variable (such as x) to that quantity. *m (for "money")*

Step 3: Translate words into numbers and symbols. This will build your equation!

First, I figured out that $150 million – $50 million – $40 million = $60 million that will earn a fee of 3%.

(I'll take the dollar sign out until I solve the equation, then put it back in if I need to in the answer.)

Next, I'll change 3% into .03

→ $60{,}000{,}000\ (.03) = m$

Step 4: Solve the equation.

$$\begin{array}{r} 60{,}000{,}000 \\ \times\ .03 \\ \hline 1{,}800{,}000.00 \end{array}$$

→ $1{,}800{,}000 = m$

So, the lawyer will earn $1.8 million at the 3% commission rate.

Step 5: Replace the variable with your answer and check that it makes the equation true.

(not really necessary in this specific instance)

Did your answer make the equation true?

- If yes: Congratulations!
- If no: Check your translation; check your math; try it again.

Step 6: How do you know your answer is reasonable?

(Answers will vary.) *1% of $60,000,000 is $600,000, so 3% would be three times as much: $1,800,000 (or $1.8 million).*

HIP-HOP MATH STUDENT CHALLENGE

Closing the Deal, Part 2

Scenario: To complete the deal, Jay-Z is thinking about hiring a high-powered attorney who will be paid on a sliding scale. Instead of a salary, the lawyer will receive a percentage of Jay-Z's total contract from the $150 million LiveNation 360 deal. The structure of the fees is as follows:

- 5% for the first $50 million
- 4% for the next $40 million
- 3% for all monies thereafter

How much money does the lawyer earn on the total deal?

How do you know your answer is reasonable?

____

____

____

HIP-HOP MATH ANSWER GUIDE

Closing the Deal, Part 2

Scenario: To complete the deal, Jay-Z is thinking about hiring a high-powered attorney who will be paid on a sliding scale. Instead of a salary, the lawyer will receive a percentage of Jay-Z's total contract from the $150 million LiveNation 360 deal. The structure of the fees is as follows:

- 5% for the first $50 million
- 4% for the next $40 million
- 3% for all monies thereafter

Close Reading Warning: Part 1 (from page 30) asks a different question than the one you are seeing here!

How much money does the lawyer earn on the total deal?

Step 1: What are you trying to find? *The amount the lawyer earns on the entire amount ($150 million).*

Step 2: Assign a variable (such as *x*) to that quantity. *t (for "total")*

Step 3: Translate words into numbers and symbols. This will build your equation!

The lawyer earns 5% on $50 million, 4% on $40 million and 3% on $60 million.

(I'll take the dollar signs out, and make all dollar amounts in millions; this will make the amounts easier to work with.)

Next, I'll change 5% to .05
4% to .04
3% to .03

Now my equation is:

$$50\,(.05) + 40\,(.04) + 60\,(.03) = t$$

Step 4: Solve the equation.

$$[50\,(.05)] + [40\,(.04)] + [60\,(.03)] = t$$

$$\rightarrow \quad [2.5 + 1.6 + 1.8] = t$$

$$\rightarrow \quad 5.9 = t$$

(Now I put the dollar sign ($) back in for the answer.)

So the lawyer will earn $5.9 million on the total deal.

Step 5: Replace the variable with your answer and check that it makes the equation true.

(Not really necessary in this specific instance, but it's always good to check with a calculator or verify the answer with someone else in the class)

Did your answer make the equation true?

If yes: Congratulations!

If no: Check your translation; check your math; try it again.

Step 6: How do you know your answer is reasonable?

(Answers will vary.) *If the lawyer were to earn an average of 4% for the entire deal, that would be $6 million—pretty close to the $5.9 million I calculated above.*

HIP-HOP MATH STUDENT CHALLENGE

Merchandise Sales

Scenario: Part of Jay-Z's 360 deal with Live Nation includes sharing the profits from merchandise sales. A T-shirt on Jay-Z's website costs $24.95. Hypothetically, if 60% of the sales price goes to cover production and marketing costs and the remaining 40% serves as profit, then how much does Jay-Z earn on the sale of each T-shirt if Live Nation keeps 80% of the profit?

How do you know your answer is reasonable?

HIP-HOP MATH ANSWER GUIDE

Merchandise Sales

Scenario: Part of Jay-Z's 360 deal with Live Nation includes sharing the profits from merchandise sales. A T-shirt on Jay-Z's website costs $24.95. Hypothetically, if 60% of the sales price goes to cover production and marketing costs and the remaining 40% serves as profit, then how much does Jay-Z earn on the sale of each T-shirt if Live Nation keeps 80% of the profit?

Step 1: What are you trying to find? *How much Jay-Z earns on each T-shirt sale (after everyone takes their share).*

Step 2: Assign a variable (such as *x*) to that quantity. *p (for "profit")*

First I need to figure out how much of the $24.95 is profit. Since the profit is 40%, I'll multiply $24.95 x .4 to get $9.98 (this is the total profit on the T-shirt).

Next, I know Live Nation keeps 80% of this, so that means Jay-Z keeps 20% of the $9.98. Now I am ready to build and solve my equation.

Step 3: Translate words into numbers and symbols. This will build your equation!

I'll take the dollar sign out and change 20% to .20

$9.98 \times .20 = p$

Step 4: Solve the equation.

$$\begin{array}{r} 9.98 \\ \times\ .20 \\ \hline 1.996 \end{array}$$

→ *This rounds up to $2 per T-shirt for Jay-Z.*

Step 5: Replace the variable with your answer and check that it makes the equation true.

(Not really necessary on this problem, but it's always good to check your math!)

Did your answer make the equation true?

If yes: Congratulations!

If no: Check your translation; check your math; try it again.

Step 6: How do you know your answer is reasonable?
(Answers will vary.) *50% of the profits would be around $5, so 20% would be less than half of that—around $2.*

HIP-HOP MATH STUDENT CHALLENGE

Real-World Scenario: Deal or No Deal? Negotiation #1

PLATINUM PROBLEM

Scenario: You are the manager for an up-and-coming independent rap artist who last year earned ...

- $8 per album (Note: Artist sold 35,000 albums.)
- $7,500 per month in digital sales
- $2,000 per month from a shoe sponsorship deal
- $6,500 per show (Note: Artist performed 10 live shows.)

A major record label comes to you and proposes a one-year 360 deal, to include ...

- $2 per album sold (Note: They guarantee at least 35,000 albums sold.)
- $5,500 per month from digital sales earnings
- $0 for shoe endorsement (Note: Artist must surrender these monies.)
- $7,000 per show for 8-date West Coast tour
- A $230,000 signing bonus, payable immediately

As the manager, the artist is looking to you to make the best business decision. **Should you accept the 360 deal?** Justify your thinking by explaining why or why not.

Work Space:

Justify your decision in 2-3 well-written sentences.

___

___

___

HIP-HOP MATH ANSWER GUIDE

Real-World Scenario: Deal or No Deal? Negotiation #1

PLATINUM PROBLEM

Scenario: You are the manager for an up-and-coming independent rap artist who last year earned …

- $8 per album (Note: Artist sold 35,000 albums.)
- $7,500 per month in digital sales
- $2,000 per month from a shoe sponsorship deal
- $6,500 per show (Note: Artist performed 10 live shows.)

A major record label comes to you and proposes a one-year 360 deal, to include …

- $2 per album sold (Note: They guarantee at least 35,000 albums sold.)
- $5,500 per month from digital sales earnings
- $0 for shoe endorsement (Note: Artist must surrender these monies.)
- $7,000 per show for 8-date West Coast tour
- A $230,000 signing bonus, payable immediately

As the manager, the artist is looking to you to make the best business decision. **Should you accept the 360 deal?** Justify your thinking by explaining why or why not.

Work Space:

Independent

*Album:
$8 per album x 35,000 albums = $280,000

*Digital Sales:
$7,500 per month x 12 months = $90,000

*Shoe Deal:
$2,000 per month x 12 months = $24,000

*Live Shows:
$6,500 per show x 10 shows = $65,000

Total: $459,000

First Offer

*Album:
$2 per album x 35,000 albums = $70,000

*Digital Sales:
$5,500 per month x 12 months = $66,000

*Shoe Deal: $0

*Live Shows:
$7,000 per show x 8 shows = $56,000

*Signing Bonus: $230,000

Total: $422,000

Justify your decision in 2-3 well-written sentances.

(Answers will vary.) I would NOT advise that the artist take the deal. As an independent artist, my client earned $459,000, which is $37,000 more than if he were to sign with the record label ($422,000).

HIP-HOP MATH STUDENT CHALLENGE

Real-World Scenario: Deal or No Deal? Negotiation #2

PLATINUM PROBLEM

Scenario: Good news! The record label is willing to negotiate, but only on one aspect of the contract that's been offered.

If you, as the manager, can get the record company to increase the amount the artist earns to $5 per album with a guarantee of 40,000 albums sold, **should the artist take this deal?** Justify your thinking by explaining why or why not.

Remain Independent:

- $8 per album (Note: Artist sold 35,000 albums.)
- $7,500 per month in digital sales
- $2,000 per month from a shoe sponsorship deal
- $6,500 per show (Note: Artist performed 10 live shows.)

360 Deal—The Counteroffer:

- $5 per album sold (Note: They guarantee at least 40,000 albums sold.)
- $5,500 per month from digital sales earnings
- $0 for shoe endorsement (Note: Artist must surrender these monies.)
- $7,000 per show for 8-date West Coast tour
- A $230,000 signing bonus, payable immediately

Work Space:

Should the artist take the deal? Justify your decision in 2-3 well-written sentences.

___

___

___

HIP-HOP MATH ANSWER GUIDE

Real-World Scenario: Deal or No Deal? Negotiation #2

PLATINUM PROBLEM

Scenario: Good news! The record label is willing to negotiate, but only on one aspect of the contract that's been offered.

If you, as the manager, can get the record company to increase the amount the artist earns to $5 per album with a guarantee of 40,000 albums sold, **should the artist take this deal?** Justify your thinking by explaining why or why not.

Remain Independent:

- $8 per album (Note: Artist sold 35,000 albums.)
- $7,500 per month in digital sales
- $2,000 per month from a shoe sponsorship deal
- $6,500 per show (Note: Artist performed 10 live shows.)

360 Deal—The Counteroffer:

- $5 per album sold (Note: They guarantee at least 40,000 albums sold.)
- $5,500 per month from digital sales earnings
- $0 for shoe endorsement (Note: Artist must surrender these monies.)
- $7,000 per show for 8-date West Coast tour
- A $230,000 signing bonus, payable immediately

Work Space:

Independent

*Album: $8 per album x 35,000 albums =	$280,000
*Digital Sales: $7,500 per month x 12 months =	$90,000
*Shoe Deal: $2,000 per month x 12 months =	$24,000
*Live Shows: $6,500 per show x 10 shows =	$65,000
Total:	$459,000

Counter Offer

*Album: $5 per album x 40,000 albums =	$200,000
*Digital Sales: $5,500 per month x 12 months =	$66,000
*Shoe Deal:	$0
*Live Shows: $7,000 per show x 8 shows =	$56,000
*Signing Bonus:	$230,000
Total:	$552,000

Should the artist take the deal? Justify your decision in 2-3 well-written sentences.

(Answers will vary.) Yes! The artist should take the deal. The record company's counteroffer will earn the artist $552,000, which is $93,000 more than what the artist earned last year ($459,000) while independent.

HIP-HOP MATH STUDENT CHALLENGE

Real-World Scenario: Deal or No Deal? Negotiation #3

PLATINUM PROBLEM

Scenario: Oh no! At the very last minute, the record label has changed the terms and made its final offer to the artist.

Remain Independent:

- $8 per album (Note: Artist sold 35,000 albums.)
- $7,500 per month in digital sales (for an entire year)
- $2,000 per month from a shoe sponsorship deal
- $6,500 per show (Note: Artist performed 10 live shows.)

360 Deal—The Final Offer:

- $5 per album sold (Note: They guarantee at least 40,000 albums sold.)
- $2,000 per month from digital sales earnings (for an entire year)
- $0 for shoe endorsement (Note: Artist must surrender these monies.)
- $7,000 per show for 8-date West Coast tour
- A $230,000 signing bonus, payable immediately

Should the manager recommend that the artist take the 360 deal or remain independent?

Work Space:

Should the artist take the final offer? Justify your decision in 2-3 well-written sentences.

HIP-HOP MATH ANSWER GUIDE

Real-World Scenario: Deal or No Deal? Negotiation #3

PLATINUM PROBLEM

Scenario: Oh no! At the very last minute, the record label has changed the terms and made its final offer to the artist.

Remain Independent:

- $8 per album (Note: Artist sold 35,000 albums.)
- $7,500 per month in digital sales (for an entire year)
- $2,000 per month from a shoe sponsorship deal
- $6,500 per show (Note: Artist performed 10 live shows.)

360 Deal—The Final Offer:

- $5 per album sold (Note: They guarantee at least 40,000 albums sold.)
- $2,000 per month from digital sales earnings (for an entire year)
- $0 for shoe endorsement (Note: Artist must surrender these monies.)
- $7,000 per show for 8-date West Coast tour
- A $230,000 signing bonus, payable immediately

Should the manager recommend that the artist take the 360 deal or remain independent?

Work Space:

Independent

*Album:
$8 per album x 35,000 albums = $280,000

*Digital Sales:
$7,500 per month x 12 months = $90,000

*Shoe Deal:
$2,000 per month x 12 months = $24,000

*Live Shows:
$6,500 per show x 10 shows = $65,000

Total: $459,000

Final Offer

*Album:
$5 per album x 40,000 albums = $200,000

*Digital Sales:
$2,000 per month x 12 months = $24,000

*Shoe Deal: $0

*Live Shows:
$7,000 per show x 8 shows = $56,000

*Signing Bonus: $230,000

Total: $510,000

Should the artist take the final offer? Justify your decision in 2-3 well-written sentences.

(**Answers will vary.**) Yes! Even though the record company lowered its offer to $2,000 per month for digital sales earnings, taking the deal will still pay $510,000, which is $51,000 more than the artist earned last year while independent ($459,000).

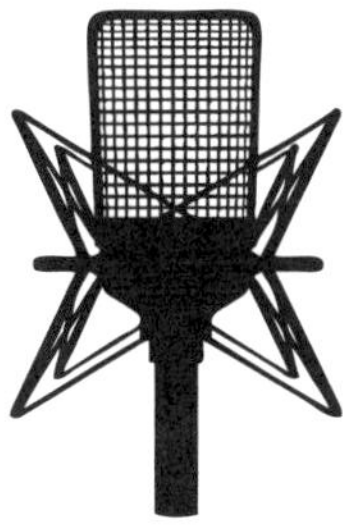

TWELVE TOP TERMS: KEY ACADEMIC VOCABULARY

Average
A single number that represents the data set.
An average is calculated by dividing the sum of the values in the set by their number.

Commission
The amount of money an employee earns when they sell something, usually a percentage of the price.

Equation
A mathematical sentence with an = sign.

Expenditure
An amount of money spent.

Invest
To give or spend money with the expectation of achieving a return.

Justify (your thinking)
To show that you are right or reasonable.

Negotiate
To attempt to reach an agreement or compromise through a discussion with others.

Percentage
A rate, number, or amount in each hundred.

Profits
A financial gain, which results from earning more than you spend.

Sliding scale
Fees that can be raised or lowered, depending on the situation.

Translate
To change from one language to another (math is a language!)

Variable
A letter that represents a quantity.

"Do not worry about your difficulties in mathematics. I can assure you mine are still greater."

—ALBERT EINSTEIN

UNIT 3

GUESS AND CHECK

Guess and check helps you:

start solving quickly ▪ eliminate wrong answers ▪ gain clarity

Step 1: Guess the answer (start with something easy and reasonable).

Step 2: Check (do the math and see how close you got).

Was your answer too high?
Was your answer too low?

Step 3: Make a new guess.

If your answer was too high, how do you need to adjust?
If your answer was too low, how do you need to adjust?

Step 4: Check again.

Are you getting closer?
Don't give up! The guess-and-check process often takes several tries.

Step 5: Find the answer.

Step 6: How do you know your answer is reasonable?

HIP-HOP MATH STUDENT CHALLENGE

Hip-Hop Fashion: Hoodie #1

Scenario: Ice Cube and Nikki Minaj are thinking about teaming up to launch a new line of hoodie sweatshirts for urban technology lovers. Some sweatshirts might have earbuds instead of drawstrings, some might have cell phone rechargers built into the sleeves—all sorts of futuristic possibilities are on the table.

To produce the base model hoodie (without any technology), a local manufacturer will charge $5 per female sweatshirt and $6 per male sweatshirt (because men's clothing requires more fabric). If they need at least 5 samples for each type of sweatshirt (i.e. they require both male and female sweatshirt samples) and pay $70 for their first total shipment of hoodies, how many male and how many female sweatshirts did they order?

Step 1: Guess the answer (start with something easy and reasonable).

Step 2: Check (do the math and see how close you got).

Step 3: Make a new guess. (Was your answer too low or too high? Adjust.)

Step 4: Check again. (Getting closer? Don't give up! Guess-and-check often takes several tries.)

Step 5: Find the answer.

Step 6: How do you know your answer is reasonable?

____

____

____

HIP-HOP MATH ANSWER GUIDE

Hip-Hop Fashion: Hoodie #1

Scenario: Ice Cube and Nikki Minaj are thinking about teaming up to launch a new line of hoodie sweatshirts for urban technology lovers. Some sweatshirts might have earbuds instead of drawstrings, some might have cell phone rechargers built into the sleeves—all sorts of futuristic possibilities are on the table.

To produce the base model hoodie (without any technology), a local manufacturer will charge $5 per female sweatshirt and $6 per male sweatshirt (because men's clothing requires more fabric). If they need at least 5 samples for each type of sweatshirt (i.e. they require both male and female sweatshirt samples) and pay $70 for their first total shipment of hoodies, how many male and how many female sweatshirts did they order?

Step 1: Guess the answer (start with something easy and reasonable).

First attempts will vary. Students may guess too high or too low.

Use this model as a "think-aloud."

Here's one example of a guess that is too high: 7 female and 7 male sweatshirts.

Step 2: Check (do the math and see how close you got).

7 female sweatshirts x $5 = $35
7 male sweatshirts x $6 = $42
Total = $77

Too high, so I probably need to decrease at least one of the numbers.

Step 3: Make a new guess.

Since my first guess was too high, I'll try 6 female and 6 male sweatshirts.

Step 4: Check again.

6 female sweatshirts x $5 = $30
6 male sweatshirts x $6 = $36
Total = $66

A little bit closer. Now my answer is too low. I'll try some different numbers until I get the answer.

Step 5: Find the answer.

8 female sweatshirts x $5 = $40
5 male sweatshirts x $6 = $30
Total = $70

8 female and 5 male sweatshirts were ordered.

Step 6: How do you know your answer is reasonable?

(Answers will vary.) *There are at least 5 samples of each type of sweatshirt, and the total cost is $70.*

HIP-HOP MATH STUDENT CHALLENGE

Hip-Hop Fashion: Hoodie #2

Scenario: Ice Cube and Nikki Minaj are thinking about teaming up to launch a new line of hoodie sweatshirts for urban technology lovers. Some sweatshirts might have earbuds instead of drawstrings, some might have cell phone rechargers build into the sleeves—all sorts of futuristic possibilities are on the table.

A different local manufacturer offers to produce the base model hoodie (without any technology) for $4 per female sweatshirt and $7 per male sweatshirt (because men's clothing requires more fabric). If they need at least 4 samples for each type of sweatshirt (i.e. they require both male and female sweatshirt samples) and pay $73 for their first total shipment of hoodies, how many male and how many female sweatshirts did they order?

Step 1: Guess the answer (start with something easy and reasonable).

Step 2: Check (do the math and see how close you got).

Step 3: Make a new guess. (Was your answer too low or too high? Adjust.)

Step 4: Check again. (Getting closer? Don't give up! Guess-and-check often takes several tries.)

Step 5: Find the answer.

Step 6: How do you know your answer is reasonable?

HIP-HOP MATH ANSWER GUIDE

Hip-Hop Fashion: Hoodie #2

Scenario: Ice Cube and Nikki Minaj are thinking about teaming up to launch a new line of hoodie sweatshirts for urban technology lovers. Some sweatshirts might have earbuds instead of drawstrings, some might have cell phone rechargers build into the sleeves—all sorts of futuristic possibilities are on the table.

A different local manufacturer offers to produce the base model hoodie (without any technology) for $4 per female sweatshirt and $7 per male sweatshirt (because men's clothing requires more fabric). If they need at least 4 samples for each type of sweatshirt (i.e. they require both male and female sweatshirt samples) and pay $73 for their first total shipment of hoodies, how many male and how many female sweatshirts did they order?

Step 1: Guess the answer (start with something easy and reasonable).

First attempts will vary. Students may guess too high or too low.

Use this model as a "think-aloud."

Here's one example of a guess that is too low: 5 female and 5 male sweatshirts.

Step 2: Check (do the math and see how close you got).

5 female sweatshirts x $4 = $20
5 male sweatshirts x $7 = $35
Total = $55

Too low, so I probably need to increase at least one of the numbers.

Step 3: Make a new guess.

Since my first guess was too low, I'll try 6 female and 8 male sweatshirts.

Step 4: Check again.

6 female sweatshirts x $4 = $24
8 male sweatshirts x $7 = $56
Total = $80

A little bit closer. Now my answer is too high. I'll try decreasing the number of male sweatshirts to 7.

Step 5: Find the answer.

6 female sweatshirts x $4 = $24
7 male sweatshirts x $7 = $49
Total = $73

6 female and 7 male sweatshirts were ordered.

Step 6: How do you know your answer is reasonable?

(Answes will vary.) *There are at least 4 samples of each type of sweatshirt, and the total cost is $73.*

HIP-HOP MATH STUDENT CHALLENGE

Hip-Hop Fashion: Manufacturing

Scenario: After choosing a manufacturer, a larger order is placed. When the hoodies arrive, there are 7 fewer male sweatshirts than female sweatshirts in the shipment. In total, they received 83 sweatshirts. How many male and how many female sweatshirts will be itemized on the packing slip?

Step 1: Guess the answer (start with something easy and reasonable).

Step 2: Check (do the math and see how close you got).

Step 3: Make a new guess. (Was your answer too low or too high? Adjust.)

Step 4: Check again. (Getting closer? Don't give up! Guess-and-check often takes several tries.)

Step 5: Find the answer.

Step 6: How do you know your answer is reasonable?

____

____

____

HIP-HOP MATH ANSWER GUIDE

Hip-Hop Fashion: Manufacturing

Scenario: After choosing a manufacturer, a larger order is placed. When the hoodies arrive, there are 7 fewer male sweatshirts than female sweatshirts in the shipment. In total, they received 83 sweatshirts. How many male and how many female sweatshirts will be itemized on the packing slip?

Step 1: Guess the answer (start with something easy and reasonable).

First attempts will vary. Students may guess too high or too low.

Use this model as a "think-aloud."

Here's one example of a possible guess, from a student who begins by selecting two amounts that add up to 83: 40 male sweatshirts and 43 female sweatshirts.

Step 2: Check (do the math and see how close you got).

40 + 43 does equal 83; however, there are only 3 fewer male sweatshirts than female. There need to be 7 fewer, so I'm off by 4.

Since I'm off by 4, I'll lower the number of male sweatshirts by 1 (to 39) and increase the number of female sweatshirts by 3 (to 46).

Step 3: Make a new guess.

39 (male sweatshirts) is seven fewer than 46 (female); however, now the total is 85. I'm over by 2 total.

Since I'm over by 2 total, I'll subtract 1 male and 1 female sweatshirt.

Step 4: Check again.

Male sweatshirts: 39 − 1 = 38

Female sweatshirts: 46 − 1 = 45

Step 5: Find the answer.

38 male sweatshirts + 45 female sweatshirts = 83; and there are 7 fewer male sweatshirts than female.

Step 6: How do you know your answer is reasonable?

***(Answers will vary.)** I read back through the problem and made sure that the answer I got (38 male and 45 female hoodies) fit what the problem asked me to solve: 7 fewer male than female, with a total of 83 hoodies.*

HIP-HOP MATH STUDENT CHALLENGE

Hip-Hop Fashion: Color Samples

Scenario: Three different color samples were sent in the 83-sweatshirt shipment: black, white, and gray. Only 17 of these sweatshirts were gray. Of the remaining sweatshirts, there were twice as many black hoodies as there were white ones. How many of each color (black, white, and gray) were included in the shipment?

Step 1: Guess the answer (start with something easy and reasonable).

Step 2: Check (do the math and see how close you got).

Step 3: Make a new guess. (Was your answer too low or too high? Adjust.)

Step 4: Check again. (Getting closer? Don't give up! Guess-and-check often takes several tries.)

Step 5: Find the answer.

Step 6: How do you know your answer is reasonable?

____

____

____

HIP-HOP MATH ANSWER GUIDE

Hip-Hop Fashion: Color Samples

Scenario: Three different color samples were sent in the 83-sweatshirt shipment: black, white, and gray. Only 17 of these sweatshirts were gray. Of the remaining sweatshirts, there were twice as many black hoodies as there were white ones. How many of each color (black, white and gray) were included in the shipment?

Step 1: Guess the answer (start with something easy and reasonable).

First attempts will vary. Students may guess too high or too low.
Use this model as a "think-aloud."

First, I know I need to subtract the 17 gray sweatshirts from the total.
83 − 17 = 66. So now I know the remaining 66 sweatshirts are black and white.

Since there are twice as many black hoodies as there are white, I'll start by guessing:
40 black and 20 white. I also know the number of black hoodies will be an even number, since there are twice as many as there are white sweatshirts.

Step 2: Check (do the math and see how close you got).

40 is twice as much as 20, which only adds up to 60; I need the total black and white sweatshirts to add up to 66.

This total is too low, so I'll increase the number of black sweatshirts to 42, and the number of white sweatshirts to 21 (since there are twice as many black sweatshirts as white).

Step 3: Make a new guess.

42 black sweatshirts + 21 white sweatshirts = 63 total. I need 66, so I'm getting very close!

Since I'm still slightly too low, I'll increase the number of black sweatshirts to 44, and the number of white sweatshirts to 22.

Step 4: Check again.

44 black sweatshirts + 22 white sweatshirts = 66 total.

Step 5: Find the answer.

Since the scenario told us 17 gray hoodies were included in the shipment, 83-17 = 66. This is the number I am looking for. 44 black sweatshirts + 22 white sweatshirts = 66; and there are twice as many black sweatshirts as there are white. The question asks for the number of black, white and gray hoodies, so:

44 black hoodies
22 white hoodies
17 gray hoodies

Step 6: How do you know your answer is reasonable?

*(**Answers will vary.**) When I add up the 44 black, 22 white, and 17 gray hoodies, the total is 83—exactly how many the original problem said there were. Additionally, there are twice as many black hoodies as there are white ones.*

HIP-HOP MATH STUDENT CHALLENGE

Hip-Hop Fashion: Zippers

Scenario: Zippers will be used on some of the new sweatshirts. One option being considered is classic silver zippers. New technology, however, would allow some of the sweatshirts to feature highly stylized magnetic zippers that require just one hand to open or close.

The designer orders 9 total boxes of zippers with which to experiment. The classic silver zippers come 7 per box. The stylized magnetic zippers are shipped in groups of 8. If the designer received 67 zippers in all, how many boxes of each type of zipper (silver and magnetic) did she receive?

How do you know your answer is reasonable?

HIP-HOP MATH ANSWER GUIDE

Hip-Hop Fashion: Zippers

Scenario: Zippers will be used on some of the new sweatshirts. One option being considered is classic silver zippers. New technology, however, would allow some of the sweatshirts to feature highly stylized magnetic zippers that require just one hand to open or close.

The designer orders 9 total boxes of zippers with which to experiment. The classic silver zippers come 7 per box. The stylized magnetic zippers are shipped in groups of 8. If the designer received 67 zippers in all, how many boxes of each type of zipper (silver and magnetic) did she receive?

Step 1: Guess the answer (start with something easy and reasonable).

First attempts will vary. Students may guess too high or too low.
Use this model as a "think-aloud."

I know there are 9 boxes total (of both the silver and the magnetic zippers), so I'll start there. I'll guess 4 boxes of silver zippers and 5 boxes of magnetic zippers.

Step 2: Check (do the math and see how close you got).

4 boxes of silver zippers x 7 zippers per box = 28 silver zippers
5 boxes of magnetic zippers x 8 zippers per box = 40 magnetic zippers

28 silver zippers + 40 magnetic zippers = 68 total zippers
Too high by just one zipper. I'll try switching the number of boxes of each.

Step 3: Make a new guess.

5 boxes of silver zippers x 7 zippers per box = 35 silver zippers
4 boxes of magnetic zippers x 8 zippers per box = 32 magnetic zippers

35 silver zippers + 32 magnetic zippers = 67 total zippers. I got it!

Step 4: Check again.

Not needed for this one.

Step 5: Find the answer.

5 boxes of silver zippers
4 boxes of magnetic zippers

Step 6: How do you know your answer is reasonable?

(Answers will vary.) *If 5 boxes of each type of zipper had been ordered:*
5 boxes x 7 silver zippers = 35 silver zippers
5 boxes x 8 magnetic zippers = 40 magnetic zippers

That would have been 75 total zippers, which is too high. I know my answer is reasonable because it came in below 75 total zippers.

HIP-HOP MATH STUDENT CHALLENGE

Hip-Hop Fashion: Zipper Variations

Scenario: Because sweatshirts come in different sizes, so do zippers. A small classic silver zipper costs $0.50, a medium classic silver zipper costs $0.75, and a large or extra-large classic silver zipper costs $1.00. The designer orders three times as many small zippers as she does medium, and then twice as many large zippers as she does small. If the designer received 40 total zippers and paid $33.00 for the entire order, how many large zippers did she purchase?

How do you know your answer is reasonable?

____

____

____

HIP-HOP MATH ANSWER GUIDE

Hip-Hop Fashion: Zipper Variations

Scenario: Because sweatshirts come in different sizes, so do zippers. A small classic silver zipper costs $0.50, a medium classic silver zipper costs $0.75, and a large or extra-large classic silver zipper costs $1.00. The designer orders three times as many small zippers as she does medium, and then twice as many large zippers as she does small. If the designer received 40 total zippers and paid $33.00 for the entire order, how many large zippers did she purchase?

Step 1: Guess the answer (start with something easy and reasonable).

First attempts will vary. Students may guess too high or too low.
Use this model as a "think-aloud."

In this problem, we have to deal with both price and the number of zippers ordered. I'll start with the number of zippers ordered and see if I can make it equal 40 total zippers.

If I guess how many medium zippers there are, I can triple it to get the number of small zippers, then proceed to the number of large zippers.

I'll start by guessing an easy number: 10 medium zippers

Step 2: Check (do the math and see how close you got).

10 medium zippers.

Since the problem tells me there are three times as many small zippers and 3 x 10 = 30, there would be 30 small zippers.

The problem also tells me there are twice as many large zippers as there are small. Since 2 x 30 = 60, there would be 60 large zippers.

10 medium zippers + 30 small zippers + 60 large zippers = 100 total zippers

Way too high! More than double the answer I'm looking for. I'll try decreasing the number of medium zippers to half the amount: 5.

Step 3: Make a new guess.

5 medium zippers.

Since the problem tells me there are three times as many small zippers and 3 x 5 = 15, there would be 15 small zippers.

The problem also tells me there are twice as many large zippers as there are small. Since 2 x 15 = 30, there would be 30 large zippers.

5 medium zippers + 15 small zippers + 30 large zippers = 50 total zippers.

I'm too high by only 10 zippers. I'm getting close. I'll try decreasing the number of medium zippers by just one, to 4.

Answer Guide Step 4
continued on next page

Step 4: Check again.

4 medium zippers.

Since the problem tells me there are three times as many small zippers and 3 x 4 = 12, there would be 12 small zippers.

The problem also tells me there are twice as many large zippers as there are small. Since 2 x 12 = 24, there would be 24 large zippers.

4 medium zippers + 12 small zippers + 24 large zippers = 40 total zippers.

I got the total number of zippers, and now I need to make sure their cost total is $33.

Step 5: Find the answer.

4 medium zippers	x $0.75 each	= $3.00
12 small zippers	x $.50 each	= $6.00
24 large zippers	x $1.00 each	= $24.00
Total:		= $33.00

So 24 large zippers were purchased.

Step 6: How do you know your answer is reasonable?

(Answers will vary.) I was not able to find any other combination of numbers that fit the description in the problem (three times more small zippers than medium, and two times more large zippers than small) that combined to make 40 total zippers. Once I figured that out, the pricing worked out exactly as it needed to.

HIP-HOP MATH STUDENT CHALLENGE

Hip-Hop Fashion: XL Sizing

Scenario: To accommodate larger-proportioned customers, more zipperless sweatshirts are ordered in sizes 2XL and 3XL (for men only).

Regular men's sizes (S, M, L, and XL) cost $7.00 per hoodie. Each 2XL sample, however, costs an extra $1.50 per item above the base sweatshirt price, and each 3XL sample costs an extra $2.25 above the base sweatshirt price. The designer orders a total of 36 sweatshirts, for the larger customers. In the shipment there are three times as many 2XL sweatshirts as there are 3XL. What is the total amount of money that this order will cost?

How do you know your answer is reasonable?

___

___

___

HIP-HOP MATH ANSWER GUIDE

Hip-Hop Fashion: XL Sizing

Scenario: To accommodate larger-proportioned customers, more zipperless sweatshirts are ordered in sizes 2XL and 3XL (for men only).

Regular men's sizes (S, M, L, and XL) cost $7.00 per hoodie. Each 2XL sample, however, costs an extra $1.50 per item above the base sweatshirt price, and each 3XL sample costs an extra $2.25 above the base sweatshirt price. The designer orders a total of 36 sweatshirts, for the larger customers. In the shipment there are three times as many 2XL sweatshirts as there are 3XL. What is the total amount of money that this order will cost?

Step 1: Guess the answer
(start with something easy and reasonable).

First attempts will vary. Students may guess too high or too low.
Use this model as a "think-aloud."

In this problem, first we have to deal with the number of 2XL and 3XL sweatshirts order, and then deal with how much they cost. I'll start with the number of sweatshirts ordered and see if I can make it equal to a total of 36. I'll move on to cost after I figure that out.

If I guess how many 3XL sweatshirts there are, I can triple it to get the number of 2XL and then add them together to see if they equal 36 total.

I'll start by guessing an easy number: 5 3XL sweatshirts.

Step 2: Check (do the math and see how close you got).

5 3XL sweatshirts

Since the problem tells me there are three times as many 2XL sweatshirts as there are 3XL and 3 x 5 = 15, there would be 15 2XL sweatshirts.

5 3XL sweatshirts + 15 2XL sweatshirts = 20 sweatshirts total

Too low. I need it to equal 36, so for my next guess I'll increase the number of 3XL sweatshirts.

Step 3: Make a new guess.

10 3XL sweatshirts

Since the problem tells me there are three times as many 2XL sweatshirts as there are 3XL and 3 x 10 = 30, there would be 30 2XL sweatshirts.

10 3XL sweatshirts + 30 2XL sweatshirts = 40 sweatshirts total.

Now I'm slightly too high. I need it to equal 36, so for my next guess, I'll decrease the number of 3XL sweatshirts by just 1, to 9.

Answer Guide Step 4
continued on next page

Step 4: Check again.

9 3XL sweatshirts.

Since the problem tells me there are three times as many 2XL sweatshirts as there are 3XL and 3 x 9 = 27, there would be 27 2XL sweatshirts.

9 3XL sweatshirts + 27 2XL sweatshirts = 36 sweatshirts total.

I got the total number of sweatshirts, and now I need to calculate their total cost.

Step 5: Find the answer.

2XL sweatshirts cost $7.00 + $1.50 each, for a total of $8.50 each

3XL sweatshirts cost $7.00 + $2.25 each, for a total of $9.25 each

$8.50 x 27 = $229.50 total for the 2XL sweatshirts

$9.25 x 9 = $83.25 total for the 3 XL sweatshirts

$312.75 total

Step 6: How do you know your answer is reasonable?

(Answers will vary.) *If I were to estimate that each sweatshirt costs $10, then 36 of them would cost $360. Since the sweatshirts are actually cheaper, and my total is $312.75, I know my answer is reasonable.*

HIP-HOP MATH STUDENT CHALLENGE

The Royalty Riddle

PLATINUM PROBLEM

Scenario: Even though Tupac Shakur passed away in 1996, revenue is still being generated by sales of his material. According to the final legal settlement in his last will and testament, there are six parties entitled to a portion of these earnings.

- The record company owns the right to make the first financial claim. It takes ⅙ of all total yearly revenues.
- After the record company subtracts its cut of the earnings, the lawyer is legally entitled to make his claim. The lawyer takes ⅕ of what is left.
- Next comes the manager. After the lawyer has taken his share, the manager is legally entitled to take ¼ of the remaining amount of money.
- After that, the beat master claims ⅓ of the remaining revenue.
- Finally, the producer takes ½ of what's left, leaving Tupac's estate with $3 million.

Question: How much total money was originally earned last year on sales of Tupac's material?

Work Space:

HIP-HOP MATH ANSWER GUIDE

The Royalty Riddle

Scenario: Even though Tupac Shakur passed away in 1996, revenue is still being generated by sales of his material. According to the final legal settlement in his last will and testament, there are six parties entitled to a portion of these earnings.

- The record company owns the right to make the first financial claim. It takes $\frac{1}{6}$ of all total yearly revenues.
- After the record company subtracts its cut of the earnings, the lawyer is legally entitled to make his claim. The lawyer takes $\frac{1}{5}$ of what is left.
- Next comes the manager. After the lawyer has taken his share, the manager is legally entitled to take $\frac{1}{4}$ of the remaining amount of money.
- After that, the beat master claims $\frac{1}{3}$ of the remaining revenue.
- Finally, the producer takes $\frac{1}{2}$ of what's left, leaving Tupac's estate with $3 million.

Question: How much total money was originally earned last year on sales of Tupac's material?

Work Space:

Using Guess and Check

First attempts will vary. Students may guess too high or too low.
Use this model as a "think-aloud."

As my first guess, I'll start with a very simple number. I know from the problem that the record company is going to begin by taking $\frac{1}{6}$ of the total, so I'll pick an easy number that can be divided by 6. I'll guess $60 million.

The record company takes $\frac{1}{6}$ of $60 million, which is $10 million.	*There is $50 million left.*
The lawyer takes $\frac{1}{5}$ of $50 million, which is $10 million.	*There is $40 million left.*
The manager takes $\frac{1}{4}$ of $40 million, which is $10 million.	*There is $30 million left.*
The beat master takes $\frac{1}{3}$ of $30 million, which is $10 million.	*There is $20 million left.*
The producer takes $\frac{1}{2}$ of $20 million, which is $10 million.	*There is $10 million left for Tupac's estate.*

This is too high, so I'll lower my starting number to $30 million and try again ...

Eventually, students will guess $18 million. The calculations are as follows:

The record company takes $\frac{1}{6}$ of $18 million, which is $3 million.	*There is $15 million left.*
The lawyer takes $\frac{1}{5}$ of $15 million, which is $3 million.	*There is $12 million left.*
The manager takes $\frac{1}{4}$ of $12 million, which is $3 million.	*There is $9 million left.*
The beat master takes $\frac{1}{3}$ of $9 million, which is $3 million.	*There is $6 million left.*
The producer takes $\frac{1}{2}$ of $6 million, which is $3 million.	*There is $3 million left for Tupac's estate.*

Since six entities each earned $3 million, the total money earned by Tupac's estate was $18 million.

Note: Having students discuss the pattern that emerges, regardless of the beginning guess, makes for a rigorous and thought-provoking conversation in class.

HIP-HOP MATH STUDENT CHALLENGE

The Royalty Riddle: Scaffolded

PLATINUM PROBLEM

Scenario: Even though Tupac Shakur passed away in 1996, revenue is still being generated by sales of his material. According to the final legal settlement in his last will and testament, there are six parties entitled to a portion of these earnings.

- The record company owns the right to make the first financial claim. It takes $\frac{1}{6}$ of all total yearly revenues.
- After the record company subtracts its cut of the earnings, the lawyer is legally entitled to make his claim. The lawyer takes $\frac{1}{5}$ of what is left.
- Next comes the manager. After the lawyer has taken his share, the manager is legally entitled to take $\frac{1}{4}$ of the remaining amount of money.
- After that, the beat master claims $\frac{1}{3}$ of the remaining revenue.
- Finally, the producer takes $\frac{1}{2}$ of what's left, leaving Tupac's estate with $3 million.

Question: How much total money was originally earned last year on sales of Tupac's material?

One of the following answers is correct. Which one and why? Show your work.

A. $24 million
B. $20 million
C. $18 million
D. $12 million
E. $8 million
F. $6 million
G. $3 million

Work Space:

HIP-HOP MATH ANSWER GUIDE

The Royalty Riddle: Scaffolded

PLATINUM PROBLEM

Scenario: Even though Tupac Shakur passed away in 1996, revenue is still being generated by sales of his material. According to the final legal settlement in his last will and testament, there are six parties entitled to a portion of these earnings.

- The record company owns the right to make the first financial claim. It takes $\frac{1}{6}$ of all total yearly revenues.
- After the record company subtracts its cut of the earnings, the lawyer is legally entitled to make his claim. The lawyer takes $\frac{1}{5}$ of what is left.
- Next comes the manager. After the lawyer has taken his share, the manager is legally entitled to take $\frac{1}{4}$ of the remaining amount of money.
- After that, the beat master claims $\frac{1}{3}$ of the remaining revenue.
- Finally, the producer takes $\frac{1}{2}$ of what's left, leaving Tupac's estate with $3 million.

Question: How much total money was originally earned last year on sales of Tupac's material?

One of the following answers is correct. Which one and why? Show your work.

A. $24 million
B. $20 million
C. $18 million
D. $12 million
E. $8 million
F. $6 million
G. $3 million

Work Space:

The correct answer is **C.** ($18 million dollars). Here's why:

The calculations are as follows:

The record company takes $\frac{1}{6}$ of $18 million, which is $3 million. There is $15 million left.
The lawyer takes $\frac{1}{5}$ of $15 million, which is $3 million. There is $12 million left.
The manager takes $\frac{1}{4}$ of $12 million, which is $3 million. There is $9 million left.
The beat master takes $\frac{1}{3}$ of $9 million, which is $3 million. There is $6 million left.
The producer takes $\frac{1}{2}$ of $6 million, which is $3 million. There is $3 million left for Tupac's estate.

TWELVE TOP TERMS: KEY ACADEMIC VOCABULARY

Approximate
Close to the actual amount, but not completely accurate or exact.

Adjust
To change something to achieve the desired result.

Base model
The beginning sample or item *before* it has been changed.

Base price
The beginning price of an item before any additional charges have been added.

Check
To examine something to determine its accuracy.

Guess
To estimate without having enough information to be sure you are correct.

Eliminate
To remove or get rid of something.

Itemize
To list the individual units or parts.

Portion
A part of a whole; an amount, section, or piece of something.

Process of trial and error
Experimenting with different methods or amounts until you find one that is successful.

Revenue
Money earned, a.k.a. "income."

Royalty
A sum of money paid to an owner for the use of that owner's property, work, or idea(s).

UNIT 4

WORK BACKWARDS

Working backwards helps you:

understand the order of events ▪ break the problem into smaller parts ▪ see the problem from a different perspective

Step 1: Identify where the problem "ends."

Step 2: Determine the sequence of events.

Step 3: Build a visual (time line, chart, staircase, etc.) to represent the sequence.

Step 4: Start at the end and plan to work one step at a time backwards toward the beginning.

What operation(s) do you need to move one step backwards?
Do the math.

Step 5: Move another step backwards toward the beginning.

What operation(s) do you need to make the next move toward the start?
Do the math.

Step 6: Repeat, moving through the middle steps until you arrive at the start of the problem.

Step 7: How do you know your answer is reasonable?

HIP-HOP MATH STUDENT CHALLENGE

The Music Video

Scenario: Producing a music video is an extremely complex process that involves many different people and decisions. To produce a video for a new song on her upcoming album, Lauryn Hill's manager hires a production company.

The production company hires a music video director to shoot somewhere in New York City. The director, of course, has to hire a location scout, and together they begin looking at several locations throughout the city. The first day, they spend 5 hours scouting different locations in the Bronx and finish at 2 p.m. At what time did they begin earlier that day?

Step 1: Identify where the problem "ends."

Step 2: Determine the sequence of events.

Step 3: Build a visual (time line, chart, staircase, etc.) to represent the sequence.

Step 4: Start at the end and plan to work one step at a time backwards toward the beginning.

- What operation(s) do you need to move one step backwards?
- Do the math.

Step 5: Move another step backwards toward the beginning.

- What operation(s) do you need to make the next move toward the start?
- Do the math.

Step 6: Repeat, moving through the middle steps until you arrive at the start of the problem.

Step 7: How do you know your answer is reasonable?

HIP-HOP MATH ANSWER GUIDE
The Music Video

Scenario: Producing a music video is an extremely complex process that involves many different people and decisions. To produce a video for a new song on her upcoming album, Lauryn Hill's manager hires a production company.

The production company hires a music video director to shoot somewhere in New York City. The director, of course, has to hire a location scout, and together they begin looking at several locations throughout the city. The first day, they spend 5 hours scouting different locations in the Bronx and finish at 2 p.m. At what time did they begin earlier that day?

Step 1: Identify where the problem "ends."

The problem "ends" at 2 p.m.

Step 2: Determine the sequence of events.

They begin, then spend 5 hours scouting, then end at 2 p.m.

Step 3: Build a visual (time line, chart, staircase, etc.) to represent the sequence.

Begin ————————————————→ (2 p.m.)

5 hours

Step 4: Start at the end and plan to work one step at a time backwards toward the beginning.

- What operation(s) do you need to move one step backwards? *Subtract 5 hours from 2 p.m.*
- Do the math.

(9 a.m.) 10 a.m. 11 a.m. 12 p.m. 1 p.m.

Begin → 1 hr → 1 hr → 1 hr → 1 hr → 1 hr (2 p.m.)

Step 5: Move another step backwards toward the beginning.

- What operation(s) do you need to make the next move toward the start?
- Do the math.

Done in step 4.

Step 6: Repeat, moving through the middle steps until you arrive at the start of the problem.

Done in step 4.

Step 7: How do you know your answer is reasonable?

(Answers will vary.) *If they start at 9 a.m., five hours later would be 2 p.m.*

HIP-HOP MATH STUDENT CHALLENGE

The Music Video: Location Scouting

Scenario: Producing a music video is an extremely complex process that involves many different people and decisions. To produce a video for a new song on her upcoming album, Lauryn Hill's manager hires a production company.

The next day, the director and location scout head to another part of New York City to look at the possibilities in Brooklyn. If they spend 4 hours scouting and finish at 2:30 p.m., at what time did they begin earlier that day?

Step 1: Identify where the problem "ends."

Step 2: Determine the sequence of events.

Step 3: Build a visual (time line, chart, staircase, etc.) to represent the sequence.

Step 4: Start at the end and plan to work one step at a time backwards toward the beginning.

- What operation(s) do you need to move one step backwards?
- Do the math.

Step 5: Move another step backwards toward the beginning.

- What operation(s) do you need to make the next move toward the start?
- Do the math.

Step 6: Repeat, moving through the middle steps until you arrive at the start of the problem.

Step 7: How do you know your answer is reasonable?

HIP-HOP MATH ANSWER GUIDE
The Music Video: Location Scouting

Scenario: Producing a music video is an extremely complex process that involves many different people and decisions. To produce a video for a new song on her upcoming album, Lauryn Hill's manager hires a production company.

The next day, the director and location scout head to another part of New York City to look at the possibilities in Brooklyn. If they spend 4 hours scouting and finish at 2:30 p.m., at what time did they begin earlier that day?

Step 1: Identify where the problem "ends."

The problem "ends" at 2:30 p.m.

Step 2: Determine the sequence of events.

They begin, then spend 4 hours scouting, then end at 2:30 p.m.

Step 3: Build a visual (time line, chart, staircase, etc.) to represent the sequence.

Begin ——————————→ (2:30 p.m.)
4 hours

Step 4: Start at the end and plan to work one step at a time backwards toward the beginning.

- What operation(s) do you need to move one step backwards? *Subtract 4 hours from 2:30 p.m.*
- Do the math.

(10:30 a.m.) 11:30 a.m. 12:30 p.m. 1:30 p.m.

Begin → 1 hr → 1 hr → 1 hr → 1 hr (2:30 p.m.)

Step 5: Move another step backwards toward the beginning.

- What operation(s) do you need to make the next move toward the start?
- Do the math.

Done in step 4.

Step 6: Repeat, moving through the middle steps until you arrive at the start of the problem.

Done in step 4.

Step 7: How do you know your answer is reasonable?

(Answers will vary.) *If they start at 10:30 a.m., four hours later would be 2:30 p.m.*

HIP-HOP MATH STUDENT CHALLENGE

The Music Video: The Dancers

Scenario: The director decides to film in front of 1520 Sedgewick Avenue in the Bronx, a location many consider to be the birthplace of hip-hop because it is where DJ Cool Herc first started it all with two turntables and a mic.

The director holds auditions for backup dancers. At the end of the first day of auditioning, he eliminates half of the dancers trying out for a role in the video. On the second day of the auditions, he eliminates half of the remaining dancers trying out. If 14 dancers now remain, how many dancers began auditions on the first day of tryouts?

Step 1: Identify where the problem "ends."

Step 2: Determine the sequence of events.

Step 3: Build a visual (time line, chart, staircase, etc.) to represent the sequence.

Step 4: Start at the end and plan to work one step at a time backwards toward the beginning.

- What operation(s) do you need to move one step backwards?
- Do the math.

Step 5: Move another step backwards toward the beginning.

- What operation(s) do you need to make the next move toward the start?
- Do the math.

Step 6: Repeat, moving through the middle steps until you arrive at the start of the problem.

Step 7: How do you know your answer is reasonable?

____

____

HIP-HOP MATH ANSWER GUIDE

The Music Video: The Dancers

Scenario: The director decides to film in front of 1520 Sedgewick Avenue in the Bronx, a location many consider to be the birthplace of hip-hop because it is where DJ Cool Herc first started it all with two turntables and a mic.

The director holds auditions for backup dancers. At the end of the first day of auditioning, he eliminates half of the dancers trying out for a role in the video. On the second day of the auditions, he eliminates half of the remaining dancers trying out. If 14 dancers now remain, how many dancers began auditions on the first day of tryouts?

Step 1: Identify where the problem "ends."

The problem "ends" with 14 dancers.

Step 2: Determine the sequence of events.

Auditions begin; half are eliminated after day 1, and half are eliminated again at end of day 2.

Step 3: Build a visual (time line, chart, staircase, etc.) to represent the sequence.

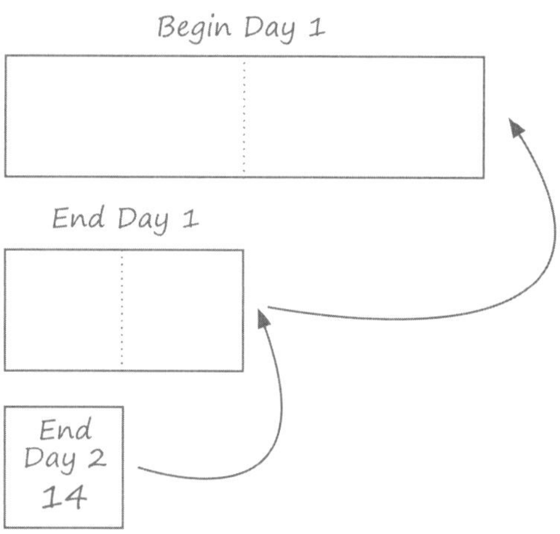

Step 4: Start at the end and plan to work one step backwards toward the beginning.

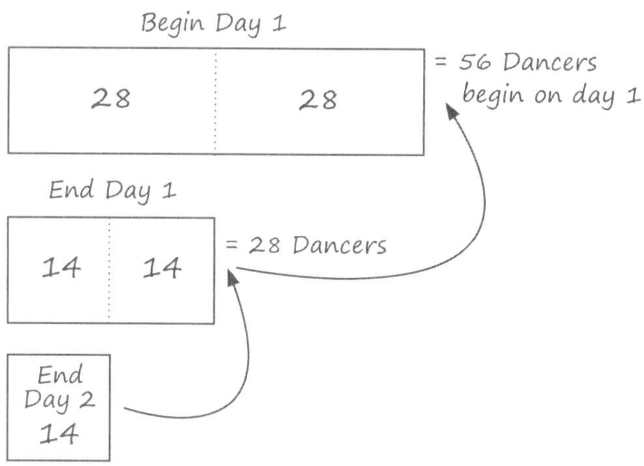

- What operation(s) do you need to move one step backwards? *First, double the 14. Then, double the 28. The answer is 56.*
- Do the math.

Step 5: Move another step backwards toward the beginning.

- What operation(s) do you need to make the next move toward the start?
- Do the math.

Done in step 4.

Step 6: Repeat, moving through the middle steps until you arrive at the start of the problem.

(Not needed in this problem.)

Step 7: How do you know your answer is reasonable?

(Answers will vary.) If the auditions start with 56 dancers and half are eliminated at the end of the first day, that's 28. If half of the 28 dancers are eliminated after the second day, 14 remain!

HIP-HOP MATH STUDENT CHALLENGE

The Music Video: Content

Scenario: The director determines that one-fourth of the final video will be focused on dancing and one-half will be focused on singing. The rest of the video will be split equally between showing buildings in the neighborhood and showing the faces of people who live there. If the total length of the video is scheduled to be 4 minutes long, how much time will be devoted to showing the faces of the area's residents?

Step 1: Identify where the problem "ends."

Step 2: Determine the sequence of events.

Step 3: Build a visual (time line, chart, staircase, etc.) to represent the sequence.

Step 4: Start at the end and plan to work one step at a time backwards toward the beginning.

- What operation(s) do you need to move one step backwards?
- Do the math.

Step 5: Move another step backwards toward the beginning.

- What operation(s) do you need to make the next move toward the start?
- Do the math.

Step 6: Repeat, moving through the middle steps until you arrive at the start of the problem.

Step 7: How do you know your answer is reasonable?

HIP-HOP MATH ANSWER GUIDE

The Music Video: Content

Scenario: The director determines that one-fourth of the final video will be focused on dancing and one-half will be focused on singing. The rest of the video will be split equally between showing buildings in the neighborhood and showing the faces of people who live there. If the total length of the video is scheduled to be 4 minutes long, how much time will be devoted to showing the faces of the area's residents?

Step 1: Identify where the problem "ends."

The problem "ends" with the video being 4 minutes long.

Step 2: Determine the sequence of events.

One-fourth of the video is dancing, one-half of the video is singing, and the rest is split equally between buildings and faces.

Step 3: Build a visual (time line, chart, staircase, etc.) to represent the sequence.

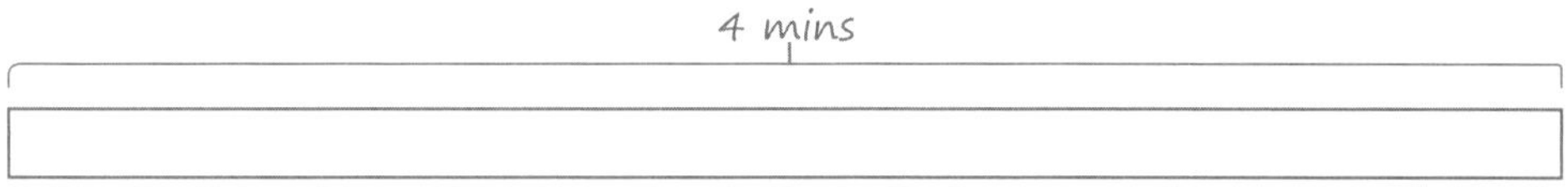

Step 4: Start at the end and plan to work one step at a time backwards toward the beginning.

- What operation(s) do you need to move one step backwards?
- Do the math.

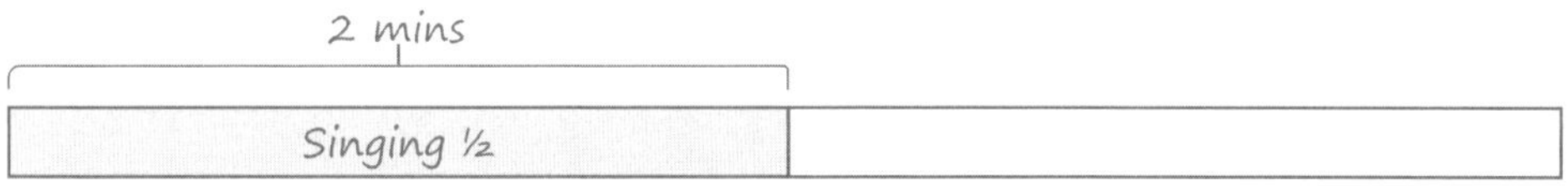

Step 5: Move another step backwards toward the beginning.

- What operation(s) do you need to make the next move towards the start?
- Do the math.

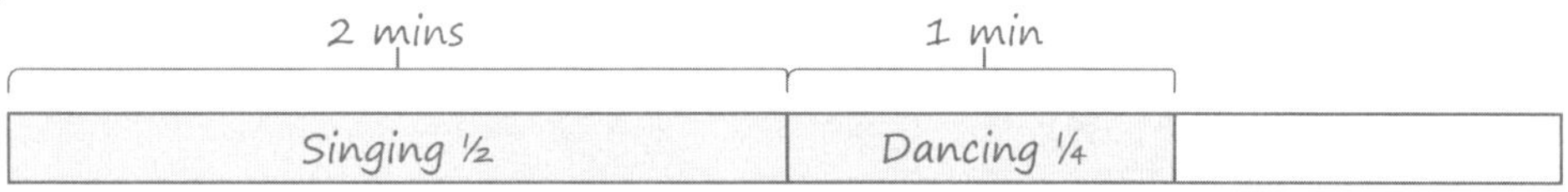

Step 6: Repeat, moving through the middle steps until you arrive at the start of the problem.

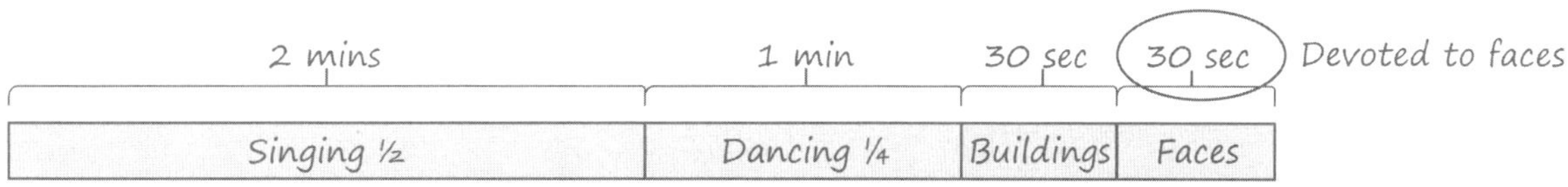

Step 7: Test your solution to see if it is reasonable.

***(Answers will vary.)** If the buildings and faces are 30 seconds each, that adds up to one minute (which is one-fourth of 4 minutes). Another minute for dancing (which is one-fourth of 4 minutes) makes 2 minutes total. The other 2 minutes are devoted to singing, which is half of the 4-minute total for the video.*

HIP-HOP MATH STUDENT CHALLENGE

The Music Video: Food and Drink

Scenario: To keep the production crew and dancers full of energy, a caterer is hired to keep the location shoot stocked with food and drinks. Bottles of iced tea were the most popular, with twice as many bottles consumed as bottles of lemonade (which was the second most popular). Soda was the least popular of the options, with one-fifth as many cans consumed as bottles of lemonade. If 17 cans of soda were consumed, how many bottles of iced tea did the video crew and dancers drink over the course of the day?

How do you know your answer is reasonable?

HIP-HOP MATH ANSWER GUIDE

The Music Video: Food and Drink

Scenario: To keep the production crew and dancers full of energy, a caterer is hired to keep the location shoot stocked with food and drinks. Bottles of iced tea were the most popular, with twice as many bottles consumed as bottles of lemonade (which was the second most popular). Soda was the least popular of the options, with one-fifth as many cans consumed as bottles of lemonade. If 17 cans of soda were consumed, how many bottles of iced tea did the video crew and dancers drink over the course of the day?

Step 1: Identify where the problem "ends."

The problem "ends" with 17 cans of sodas having been consumed.

Step 2: Determine the sequence of events.

Twice as many iced teas (most popular) were consumed as lemonade.
Lemonade was second most popular, with 5 times as many consumed as soda.
Soda was least popular, with 17 cans.

Step 3: Build a visual (time line, chart, staircase, etc.) to represent the sequence.

Step 4: Start at the end and plan to work one step at a time backwards toward the beginning.

- What operation(s) do you need to move one step backwards?
- Do the math.

Step 5: Move another step backwards toward the beginning.

- What operation(s) do you need to make the next move toward the start?
- Do the math.

Step 6: Repeat, moving through the middle steps until you arrive at the start of the problem.

Step 7: How do you know your answer is reasonable?

(Answers will vary.) If 170 iced teas were consumed, that's twice as many as 85 lemonades. One-fifth of 85 is 17, which is the number of cans of soda that were consumed.

HIP-HOP MATH STUDENT CHALLENGE

The Music Video: Unions and Overtime

Scenario: Regulations dictate that members of the film crew must be paid 1.5 times the amount of their regular wage for any overtime they work on the video shoot. An official workday is 8 hours. (Note: This 8-hour workday includes a paid 1-hour lunch break.) There are 7 people on the film crew who have been hired for the day's video shoot, and all of them will start and stop their workday at the same time.

If $3,500 was paid out in total wages to the entire film crew and the regular wage earned was $40 per hour per worker, how many hours did each person on the crew work that day?

How do you know your answer is reasonable?

____

____

____

HIP-HOP MATH ANSWER GUIDE

The Music Video: Unions and Overtime

Scenario: Regulations dictate that members of the film crew must be paid 1.5 times the amount of their regular wage for any overtime they work on the video shoot. An official workday is 8 hours. (Note: This 8-hour workday includes a paid 1-hour lunch break.) There are 7 people on the film crew who have been hired for the day's video shoot, and all of them will start and stop their workday at the same time.

If $3,500 was paid out in total wages to the entire film crew and the regular wage earned was $40 per hour per worker, how many hours did each person on the crew work that day?

Step 1: Identify where the problem "ends."

The problem "ends" with $3,500 being paid out to the film crew.

Step 2: Determine the sequence of events.

7 people on the film crew work 8 hours each and are paid $40 per hour (this is $320 per person).
The 7 people on the film crew work overtime for 1.5 times their normal rate (this is $60 per hour, or $420 per hour for all 7 people).
All 7 people are paid a total of $3,500 to the entire group.

Step 3: Build a visual (time line, chart, staircase, etc.) to represent the sequence.

Regular Workday	+	Overtime	=	$3,500

Step 4: Start at the end and plan to work one step at a time backwards toward the beginning.

- What operation(s) do you need to move one step backwards?
- Do the math.

Regular Workday	+	Overtime	=	$3,500

7 people x 8 hours = 56 hours
56 hours x $40/hour = $2,240. This means there is $3,500 – $2,240 left for overtime which equals $1,260.

Step 5: Move another step backwards toward the beginning.

- What operation(s) do you need to make the next move toward the start?
- Do the math.

Regular Workday	+	Overtime	=	$3,500
$2,240	+	$1,260	=	$3,500

$1,260 ÷ 7 people = $180 per person in overtimes
$180 ÷ $60/hour = 3 hours of overtime per worker

Step 6: Repeat, moving through the middle steps until you arrive at the start of the problem.

3 hours of overtime + 8 hours regular time = 11 hours worked (per person)

Step 7: How do you know your answer is reasonable?

(Answers will vary.)

Regular Day:
Each worker earned $40/hour x 8 hours = $320 per person (regular time)
$320 x 7 people = $2,240 (total regular time paid)

Overtime:
Each worker earned $60/hour x 3 hours = $180 per person (overtime)
$180 x 7 people = $1,260 (total overtime paid)
$2,240 (regular pay) + $1,260 (overtime) = $3,500 total paid out to the video crew

HIP-HOP MATH STUDENT CHALLENGE

The Music Video: Paying for the Permit

Scenario: To film on Sedgewick Avenue, the video production company must pay for a permit. This permit costs \$300. Additionally, the permit states that the crew must be done filming by 3:00 p.m. or the company will be fined \$55 for every 20 minutes that filming continues past that hour. If the production company ends up having to pay a total of \$740 (which includes the permit cost plus the overtime fee), what time did filming finish?

How do you know your answer is reasonable?

____

____

____

HIP-HOP MATH ANSWER GUIDE

The Music Video: Paying for the Permit

Scenario: To film on Sedgewick Avenue, the video production company must pay for a permit. This permit costs $300. Additionally, the permit states that the crew must be done filming by 3:00 p.m. or the company will be fined $55 for every 20 minutes that filming continues past that hour. If the production company ends up having to pay a total of $740 (which includes the permit cost plus the overtime fee), what time did filming finish?

Step 1: Identify where the problem "ends."

The problem "ends" with the production company paying $740 for the permit + overtime fee.

Step 2: Determine the sequence of events.

Company pays $300 for the permit. Filming must be done by 3 p.m.
Company films longer than 3 p.m. and has to pay $55 for every 20 minutes extra.
Company pays a total of $740.

Step 3: Build a visual (time line, chart, staircase, etc.) to represent the sequence.

Step 4: Start at the end and plan to work one step at a time backwards toward the beginning.

- What operation(s) do you need to move one step backwards?
- Do the math.

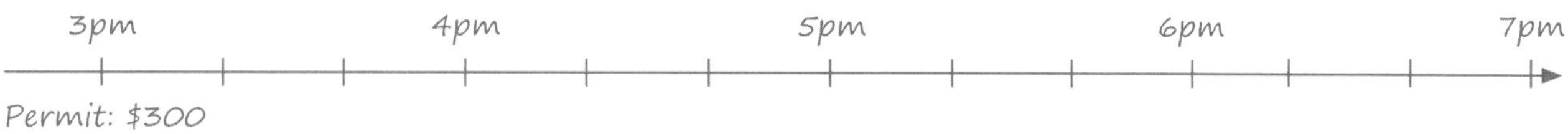

Step 5: Move another step backwards toward the beginning.

- What operation(s) do you need to make the next move toward the start?
- Do the math.

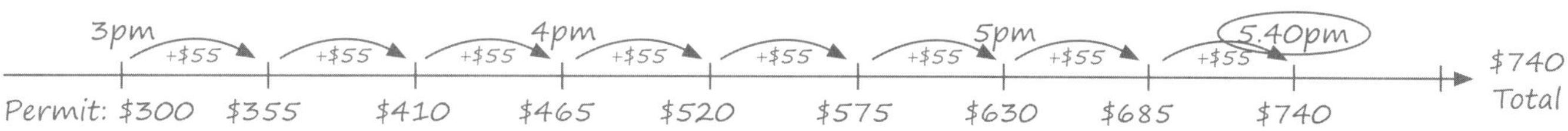

Step 6: Repeat, moving through the middle steps until you arrive at the start of the problem.

Not necessary on this problem.

Step 7: How do you know your answer is reasonable?

(Answers will vary.) *From 3:00 p.m. to 5:40 p.m., there are 8 segments of 20 minutes. 8 x $55 = $440. This is reasonable because the starting cost of the permit is $300 + $440 in overtime = $740 total. Note to teacher: Though this is not a classic "work backwards" problem, using a timeline to flow back-and-forth is often an effective skill for conquering word problems.*

HIP-HOP MATH STUDENT CHALLENGE

Make a Time Line of the Video Shoot

PLATINUM PROBLEM

Scenario: *Use the following information to create a time line of the day's events:*

(Note: These events are not listed in any particular order.)

- The video crew arrives to begin setting up at 8 a.m.
- The catering company arrives with food and drinks 2 hours after the backup dancers arrive.
- The permit says filming should be finished by 3:00 p.m.
- The backup dancers arrive 1.5 hours after the video crew arrives.
- The catering company brings a second delivery of more food and drinks 2.5 hours after its first delivery.
- The filming actually ends 2 hours and 40 minutes after the permit allows (Note: all penalty fees already accounted for).
- The director buys everyone dinner to celebrate the end of filming 11 hours after the video crew first arrived to set up.

HIP-HOP MATH STUDENT CHALLENGE

Time Line Questions

PLATINUM PROBLEM

1. How much time elapsed between the caterer's first and second deliveries?

2. How much time elapsed between when the backup dancers arrived and when the caterer made its second delivery?

3. Which was longer: the time between when the video crew arrived and 4:00 p.m., or the time between when the backup dancers arrived and when filming ended?

4. Now write your own question that can be answered using the timeline.

HIP-HOP MATH ANSWER GUIDE

Make a Time Line of the Video Shoot

PLATINUM PROBLEM

Scenario: *Use the following information to create a time line of the day's events:*

(Note: These events are not listed in any particular order.)

- The video crew arrives to begin setting up at 8 a.m.
- The catering company arrives with food and drinks 2 hours after the backup dancers arrive.
- The permit says filming should be finished by 3:00 p.m.
- The backup dancers arrive 1.5 hours after the video crew arrives.
- The catering company brings a second delivery of more food and drinks 2.5 hours after its first delivery.
- The filming actually ends 2 hours and 40 minutes after the permit allows (Note: all penalty fees already accounted for).
- The director buys everyone dinner to celebrate the end of filming 11 hours after the video crew first arrived to set up.

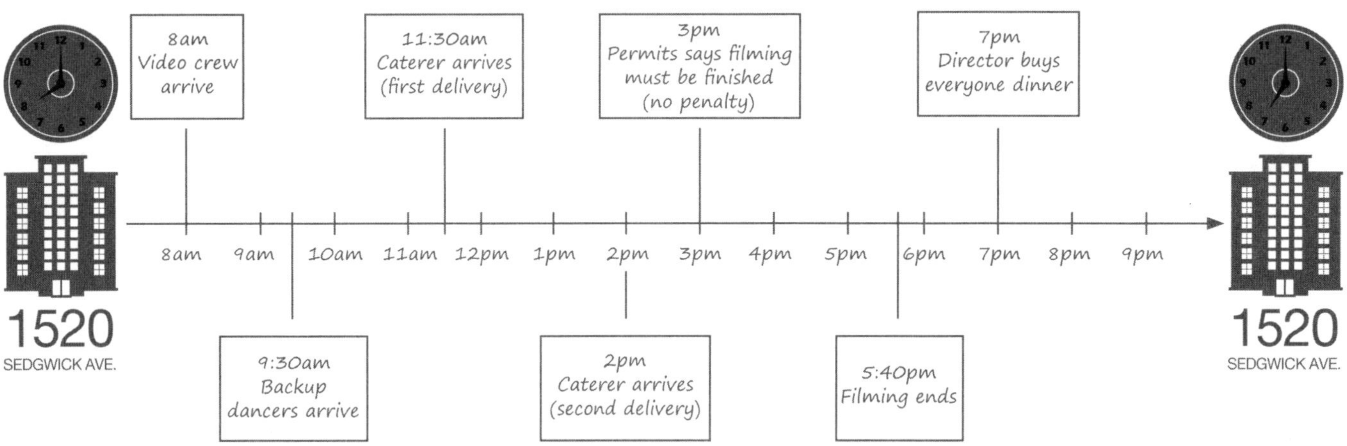

HIP-HOP MATH ANSWER GUIDE

Time Line Questions

PLATINUM PROBLEM

1. How much time elapsed between the caterer's first and second deliveries?

   2.5 hours, or 2 hours and 30 minutes.

2. How much time elapsed between when the backup dancers arrived and when the caterer made its second delivery?

   4.5 hours, or 4 hours and 30 minutes.

3. Which was longer: the time between when the video crew arrived and 4:00 p.m., or the time between when the backup dancers arrived and when filming ended?

   The time between when the video crew arrived (8:00 a.m.) and 4:00 p.m. was 8 hours.

   The time between when the backup dancers arrived (9:30 a.m.) and when filming ended (5:40 p.m.) was 8 hours and 10 minutes, so this period was 10 minutes longer.

4. Now write your own question that can be answered using the timeline.

   (Answers will vary.)

TWELVE TOP TERMS: KEY ACADEMIC VOCABULARY

Eliminate
To completely remove or get rid of.

Fine
Money paid when an agreement is broken.

Identify
To establish who or what something is.

Length
The measure of something from end to end.

Operation
An action or procedure that produces a new value.

Options
Choices or possibilities to choose from.

Overtime
Time worked in addition to normal working hours.

Permit
An official document giving someone authorization to do something.

Sequence
A particular order in which related events happen.

Time line
A graphic representation of the passage of time (shown on a line).

Twice
Two times; on two occasions.

Wage
A fixed, regular payment made by an employer to an employee.

UNIT 5

LOOK FOR A PATTERN

Looking for a pattern helps you:

make predictions ▪ see relationships ▪ identify repeating designs

Step 1: Write down all the numbers that are given. Leave some space between them.

Are the numbers increasing or decreasing?

Step 2: Look at the first two numbers. Ask yourself, "How did they get from the first amount to the second?" Write your answer between the two numbers.

Step 3: Look at the second and third numbers. Ask yourself, "How did they get from the second amount to the third?" Write your answer between the two numbers.

Step 4: Compare what you've written. Do you see a pattern forming?

Step 5: One more time—repeat for the third and fourth number (two numbers don't make a pattern; you need at least three).

Compare what you've written. Do you see a pattern yet?

Yes: Go to step 6.

No: Repeat the process until you see the pattern.

Step 6: Use the pattern to answer the question.

Step 7: How do you know your answer is reasonable?

HIP-HOP MATH STUDENT CHALLENGE

Building a Fan Base

Scenario: Kendrick Lamar's social media director has been put in charge of orchestrating a new online campaign for the artist's forthcoming album.

Based on previous data and the artist's current popularity on the charts, the social media manager anticipates the following numbers of total YouTube views:

Anticipated Total YouTube Views: Kendrick Lamar's New Video

Day 1: 9 million ▪ *Day 2*: 10 million ▪ *Day 3*: 12 million ▪ *Day 4*: 15 million

If this pattern of growth continues at the same rate, what will be the total number of YouTube views on day 5?

Step 1: Write down all the numbers that are given. Leave some space between them. Are the numbers increasing or decreasing?

Step 2: Look at the first two numbers. Ask yourself, "How did they get from the first amount to the second?" Write your answer between the two numbers.

Step 3: Look at the second and third numbers. Ask yourself, "How did they get from the second amount to the third?" Write your answer between the two numbers.

Step 4: Compare what you've written. Do you see a pattern forming?

Step 5: One more time—repeat for the third and fourth number (two numbers don't make a pattern; you need at least three).

- Do you see a pattern yet?

Step 6: Use the pattern to answer the question.

Step 7: How do you know your answer is reasonable?

HIP-HOP MATH ANSWER GUIDE

Building a Fan Base

Scenario: Kendrick Lamar's social media director has been put in charge of orchestrating a new online campaign for the artist's forthcoming album.

Based on previous data and the artist's current popularity on the charts, the social media manager anticipates the following numbers of total YouTube views:

Anticipated total YouTube Views: Kendrick Lamar's New Video

Day 1: 9 million ▪ *Day 2*: 10 million ▪ *Day 3*: 12 million ▪ *Day 4*: 15 million

If this pattern of growth continues at the same rate, what will be the total number of YouTube views on day 5?

Step 1: Write down all the numbers that are given. Leave some space between them. Are the numbers increasing or decreasing?

	Day 1	Day 2	Day 3	Day 4	
(All numbers in millions)	*9*	*10*	*12*	*15*	*(The numbers are increasing)*

Step 2: Look at the first two numbers. Ask yourself, "How did they get from the first amount to the second?" Write your answer between the two numbers.

To get from 9 to 10, 1 is added.

Day 1 Day 2 Day 3 Day 4
9 10 12 15
+1

Step 3: Look at the second and third numbers. Ask yourself, "How did they get from the second amount to the third?" Write your answer between the two numbers.

To get from 10 to 12, 2 is added.

Day 1 Day 2 Day 3 Day 4
9 10 12 15
+1 +2

Step 4: Compare what you've written. Do you see a pattern forming?

It looks like the change increases by 1 each time.

Step 5: One more time—repeat for the third and fourth number (two numbers don't make a pattern; you need at least three).

- Do you see a pattern yet?

To get from 12 to 15, 3 is added.

Day 1 Day 2 Day 3 Day 4
9 10 12 15 19
+1 +2 +3 +4

Yes, I can confirm that the change increases by 1 each time.

Step 6: Use the pattern to answer the question.

The next step would be to add 4, so 15 + 4 = 19, so the answer is 19 million views on day 5.

Step 7: How do you know your answer is reasonable?

(Answers will vary.) This answer is reasonable because if I were to start at 9 million and add 2 million views per day, that would be 17 million views after day 5, which is pretty close to the answer I got of 19 million.

HIP-HOP MATH STUDENT CHALLENGE

Building a Fan Base: Total Retweets

Scenario: Kendrick Lamar's social media director has been put in charge of orchestrating a new online campaign for the artist's forthcoming album.

Based on previous data and the artist's current popularity on the charts, the social media manager anticipates the following numbers of total Twitter retweets alerting fans to Kendrick Lamar's new video:

Anticipated Total Twitter Retweets: Kendrick Lamar's New Video

Day 1: 2 million ▪ *Day 2*: 4 million ▪ *Day 3*: 7 million ▪ *Day 4*: 11 million

If this pattern of growth continues at the same rate, what will the total number of Twitter retweets be on day 5?

Step 1: Write down all the numbers that are given. Leave some space between them. Are the numbers increasing or decreasing?

Step 2: Look at the first two numbers. Ask yourself, "How did they get from the first amount to the second?" Write your answer between the two numbers.

Step 3: Look at the second and third numbers. Ask yourself, "How did they get from the second amount to the third?" Write your answer between the two numbers.

Step 4: Compare what you've written. Do you see a pattern forming?

Step 5: One more time—repeat for the third and fourth number (two numbers don't make a pattern; you need at least three).

- Do you see a pattern yet?

Step 6: Use the pattern to answer the question.

Step 7: How do you know your answer is reasonable?

____

____

____

HIP-HOP MATH ANSWER GUIDE

Building a Fan Base: Total Retweets

Scenario: Kendrick Lamar's social media director has been put in charge of orchestrating a new online campaign for the artist's forthcoming album.

Based on previous data and the artist's current popularity on the charts, the social media manager anticipates the following numbers of total Twitter retweets alerting fans to Kendrick Lamar's new video:

Anticipated Total Twitter Retweets: Kendrick Lamar's New Video

Day 1: 2 million ▪ *Day 2*: 4 million ▪ *Day 3*: 7 million ▪ *Day 4*: 11 million

If this pattern of growth continues at the same rate, what will be the total number of Twitter retweets on day 5?

Step 1: Write down all the numbers that are given. Leave some space between them. Are the numbers increasing or decreasing?

	Day 1	Day 2	Day 3	Day 4	
(All numbers in millions)	2	4	7	11	*(The numbers are increasing)*

Step 2: Look at the first two numbers. Ask yourself, "How did they get from the first amount to the second?" Write your answer between the two numbers.

	Day 1	Day 2	Day 3	Day 4
To get from 2 to 4, 2 is added.	2	4	7	11
	+2			

Step 3: Look at the second and third numbers. Ask yourself, "How did they get from the second amount to the third?" Write your answer between the two numbers.

	Day 1	Day 2	Day 3	Day 4
To get from 4 to 7, 3 is added.	2	4	7	11
	+2	+3		

Step 4: Compare what you've written. Do you see a pattern forming?

It looks like the change increases first by 2, then by 1 more each time.

Step 5: One more time—repeat for the third and fourth number (two numbers don't make a pattern; you need at least three).

- Do you see a pattern yet?

	Day 1	Day 2	Day 3	Day 4	
To get from 7 to 11, 4 is added.	2	4	7	11	16
	+2	+3	+4	+5	

Yes, I can confirm that the change increases by 2 each time each time.

Step 6: Use the pattern to answer the question.

The next step would be to add 5, so 11 + 5 = 16, so the answer is 16 million retweets on day 5.

Step 7: How do you know your answer is reasonable?

(Answers will vary.) This answer is reasonable because if I were to start at 2 million and add 3 million retweets per day, that would be 14 million retweets after day 5, which is pretty close to the answer I got of 16 million.

HIP-HOP MATH STUDENT CHALLENGE

Building a Fan Base: Anticipated Facebook Likes

Scenario: On Facebook, the social media manager anticipates the following numbers of total Likes for Kendrick Lamar's new video:

Anticipated Facebook Likes: Kendrick Lamar's New Video

Day 1: 1.5 million ▪ *Day 2*: 3 million ▪ *Day 3*: 5 million ▪ *Day 4*: 7.5 million ▪ *Day 5*: 10.5 million

If this pattern of growth continues at the same rate, what will be the number of Facebook Likes on day 8?

Step 1: Write down all the numbers that are given. Leave some space between them. Are the numbers increasing or decreasing?

Step 2: Look at the first two numbers. Ask yourself, "How did they get from the first amount to the second?" Write your answer between the two numbers.

Step 3: Look at the second and third numbers. Ask yourself, "How did they get from the second amount to the third?" Write your answer between the two numbers.

Step 4: Compare what you've written. Do you see a pattern forming?

Step 5: One more time—repeat for the third and fourth number (two numbers don't make a pattern; you need at least three).

- Do you see a pattern yet?

Step 6: Use the pattern to answer the question.

Step 7: How do you know your answer is reasonable?

HIP-HOP MATH ANSWER GUIDE

Building a Fan Base: Anticipated Facebook Likes

Scenario: On Facebook, the social media manager anticipates the following numbers of total Likes for Kendrick Lamar's new video:

Anticipated FaceBook Likes: Kendrick Lamar's New Video

Day 1: 1.5 million ▪ *Day 2*: 3 million ▪ *Day 3*: 5 million ▪ *Day 4*: 7.5 million ▪ *Day 5*: 10.5 million

If this pattern of growth continues at the same rate, what will be the number of Facebook Likes on day 8?

Step 1: Write down all the numbers that are given. Leave some space between them. Are the numbers increasing or decreasing?

	Day 1	Day 2	Day 3	Day 4	Day 5	
(All numbers in millions)	*1.5*	*3*	*5*	*7.5*	*10.5*	*(The numbers are increasing)*

Step 2: Look at the first two numbers. Ask yourself, "How did they get from the first amount to the second?" Write your answer between the two numbers.

	Day 1	Day 2	Day 3	Day 4	Day 5
To get from 1.5 to 3, 1.5 is added.	*1.5*	*3*	*5*	*7.5*	*10.5*

+1.5

Step 3: Look at the second and third numbers. Ask yourself, "How did they get from the second amount to the third?" Write your answer between the two numbers.

	Day 1	Day 2	Day 3	Day 4	Day 5
To get from 3 to 5, 2 is added.	*1.5*	*3*	*5*	*7.5*	*10.5*

+1.5 *+2*

Step 4: Compare what you've written. Do you see a pattern forming?

It looks like the change increases by 0.5 each time.

Step 5: One more time—repeat for the third and fourth number (two numbers don't make a pattern; you need at least three).

- Do you see a pattern yet?

	Day 1	Day 2	Day 3	Day 4	Day 5
To get from 5 to 7.5, 2.5 is added.	*1.5*	*3*	*5*	*7.5*	*10.5*

+1.5 *+2* *+2.5*

Yes, I can confirm that the change increases by 0.5 each time.

Step 6: Use the pattern to answer the question.

On day 4, there are 7.5 Likes, so I'll add ***3.0*** *to get 10.5.*
On day 5, there are 10.5 Likes, so I'll add ***3.5*** *to get 14.0.*
On day 6, there are 14.0 Likes, so I'll add ***4.0*** *to get 18.0.*
On day 7, there are 18.0 Likes, so I'll add ***4.5*** *to get 22.5.*
So the number of Likes will be 22.5 million on day 8.

Step 7: How do you know your answer is reasonable?

(Answers will vary.) *This answer is reasonable because if I were to start at 1.5 million and add 3 million Likes per day, that would be 22.5 million Likes after day 8, which is the answer I got of 22.5 million.*

HIP-HOP MATH STUDENT CHALLENGE

Building a Fan Base: Breaking 30 Million

Scenario: If the pattern of total Facebook Likes continues at the same rate in the previous scenario, on what day will the total number of Likes for Kendrick Lamar's new video surpass 30 million?

How do you know your answer is reasonable?

____

____

____

HIP-HOP MATH ANSWER GUIDE

Building a Fan Base: Breaking 30 Million

Scenario: If the pattern of total Facebook Likes continues at the same rate in the previous scenario, on what day will the total number of Likes for Kendrick Lamar's new video surpass 30 million?

In keeping with this pattern:

Day	# of Likes	Change
1	1.5 million	
		+1.5 million
2	3 million	
		+2.0 million
3	5 million	
		+2.5 million
4	7.5 million	
		+3.0 million
5	10.5 million	
		+3.5 million
6	14 million	
		+4 million
7	18 million	
		+4.5 million
8	22.5 million	
		+5 million
(9)	27.5 million	
		+5.5 million
10	33 million	

Day 9 starts with 27.5 million and ends with 33 million. Therefore, Kendrick Lamar's new video will surpass 30 million views on Day 9.

Step 7: How do you know your answer is reasonable?

(Answers will vary.) If the number of likes averaged 3 million per day for 10 days, that would be 30 million, which is pretty close to 27.5 million after 9 days."

HIP-HOP MATH STUDENT CHALLENGE

Building a Fan Base: Instagram First Look

Scenario: The social media manager decides to offer an exclusive first look at Kendrick Lamar's new album cover to Instagram users.

- On the first day, there were 100,000 favorable new comments.
- On the second day, there were 90,000 favorable new comments.
- On the third day, there were 75,000 favorable new comments.
- On the fourth day, there were 55,000 favorable new comments.

If this pattern continues, on which day will there be fewer than 15,000 favorable new comments posted per day on Instagram?

How do you know your answer is reasonable?

____

____

____

HIP-HOP MATH ANSWER GUIDE

Building a Fan Base: Instagram First Look

Scenario: The social media manager decides to offer an exclusive first look at Kendrick Lamar's new album cover to Instagram users.

- On the first day, there were 100,000 favorable new comments.
- On the second day, there were 90,000 favorable new comments.
- On the third day, there were 75,000 favorable new comments.
- On the fourth day, there were 55,000 favorable new comments.

If this pattern continues, on which day will there be fewer than 15,000 favorable new comments posted per day on Instagram?

Step 1: Write down all the numbers that are given. Leave some space between them. Are the numbers increasing or decreasing?

	Day 1	Day 2	Day 3	Day 4	
(All numbers in thousands)	*100*	*90*	*75*	*55*	*(The numbers are decreasing)*

Step 2: Look at the first two numbers. Ask yourself, "How did they get from the first amount to the second?" Write your answer between the two numbers.

To get from 100 to 90, 10 is subtracted (second day).

Day 1		Day 2	Day 3	Day 4
100		90	75	55
	−10			

Step 3: Look at the second and third numbers. Ask yourself, "How did they get from the second amount to the third?" Write your answer between the two numbers.

To get from 90 to 75, 15 is subtracted (third day).

Day 1		Day 2		Day 3	Day 4
100		90		75	55
	−10		−15		

Step 4: Compare what you've written. Do you see a pattern forming?

Not really sure yet, so I'll keep going ...

Step 5: One more time—repeat for the third and fourth number (two numbers don't make a pattern; you need at least three).

- Do you see a pattern yet?

To get from 75 to 55, 20 is subtracted (fourth day).

Day 1		Day 2		Day 3		Day 4
100		90		75		55
	−10		−15		−20	

Yes, I can confirm that the change decreases by 5 each time. So even though the number of new comments is decreasing each day, the amount of the change is actually getting larger each day.

Answer Guide Step 6 continued on next page

Step 6: Use the pattern to answer the question.

In keeping with this pattern:

Day	# of Favorable Comments (in thousands)	Change
1	100	
		−10
2	90	
		−15
3	75	
		−20
4	55	
		−25
5	30	
		−30
6	0	

Day 6: Most likely, there would not be exactly 0 new comments on Day 6. However, according to this pattern, there will be fewer than 15,000 favorable new comments posted on Day 6.

Step 7: How do you know your answer is reasonable?

(Answers will vary.) *Anything past 6 days would be a negative number, and you can't have a negative number of new comments, so Day 6 is the only possible answer. According to this pattern, it can't go any further.*

HIP-HOP MATH STUDENT CHALLENGE

Building a Fan Base: Relationship To Downloads

Scenario: The social media manager discovers there is a relationship between the number of Instagram comments and the number of downloads from iTunes.

- For the first 10,000 comments, there were 1,000 downloads.
- For the next 10,000 comments, there were 930 downloads.
- For the next 10,000 comments, there were 855 downloads.
- For the next 10,000 comments, there were 775 downloads.
- For the next 10,000 comments, there were 690 downloads.

If this pattern continues at the same rate, how many total downloads of Kendrick Lamar's new album will there be once there are 70,000 comments on Instagram?

How do you know your answer is reasonable?

HIP-HOP MATH ANSWER GUIDE

Building a Fan Base: Relationship To Downloads

Scenario: The social media manager discovers there is a relationship between the number of Instagram comments and the number of downloads from iTunes.

- For the first 10,000 comments, there were 1,000 downloads.
- For the next 10,000 comments, there were 930 downloads.
- For the next 10,000 comments, there were 855 downloads.
- For the next 10,000 comments, there were 775 downloads.
- For the next 10,000 comments, there were 690 downloads.

If this pattern continues at the same rate, how many total downloads of Kendrick Lamar's new album will there be once there are 70,000 comments on Instagram?

Step 1: Write down all the numbers that are given. Leave some space between them. Are the numbers increasing or decreasing?

1,000 930 855 755 690 (The numbers of downloads are decreasing)

-70 -75 -80

Step 2: Look at the first two numbers. Ask yourself, "How did they get from the first amount to the second?" Write your answer between the two numbers.

To get from 1,000 to 930, 70 is subtracted (second day).

Step 3: Look at the second and third numbers. Ask yourself, "How did they get from the second amount to the third?" Write your answer between the two numbers.

To get from 930 to 855, 75 is subtracted (third day).

Step 4: Compare what you've written. Do you see a pattern forming?

Not really sure yet, so I'll keep going ...

Step 5: One more time—repeat for the third and fourth number (two numbers don't make a pattern; you need at least three).

- Do you see a pattern yet?

To get from 855 to 775, 80 is subtracted (fourth day).

Yes, I see that for each new 10,000 comments, the change in the number of downloads decreases by 5 each time. So even though the number of downloads decreases with each 10,000 comments, the amount of the change is actually getting larger each day.

Answer Guide Step 6 continued on next page

Step 6: Use the pattern to answer the question.

In keeping with this pattern:

Comments on Instagram	Total Comments	# of iTunes Downloads	Change	Total Downloads
First 10,000	10,000	1,000		1,000
			−70	
Next 10,000	20,000	930		1,930
			−75	
Next 10,000	30,000	855		2,785
			−80	
Next 10,000	40,000	775		3,560
			−85	
Next 10,000	50,000	690		4,250
			−90	
Next 10,000	60,000	600		4,850
			−95	
Next 10,000	70,000	505		5,355

So, after 70,000 Instagram comments, there will be 5,355 downloads on iTunes.

Step 7: How do you know your answer is reasonable?

(Answers will vary.) *In the first part of this problem, there are 10,000 comments for 1,000 downloads; that's about 10 downloads for every 1 comment. By the end, there are 70,000 comments, which would be around 7,000 downloads. 5,355 is pretty close to 7,000 and is still less than 7,000.*

HIP-HOP MATH STUDENT CHALLENGE

Building a Fan Base: One Million Subscribers

Scenario: Kendrick Lamar's social media manager wants to build a newsletter list that exceeds 1 million subscribers. Currently, Kendrick Lamar has 70,000 people on his subscription list.

To hit her goal, the social media manager decides to run a new promotion. All fans who sign up for Kendrick Lamar's online newsletter will receive one free ringtone.

The results of the campaign demonstrate a pattern:

- On day 1, 180,000 new subscribers joined the list.
- On day 2, 172,500 new subscribers joined the list.
- On day 3, 160,000 new subscribers joined the list.
- On day 4, 142,500 new subscribers joined the list.
- On day 5, 120,000 new subscribers joined the list.

If day 1 of the campaign started on a Friday and the pattern of new subscribers to Kendrick Lamar's e-mail list continues at the same rate, on what day of the week will the social media manager reach her goal of exceeding 1 million subscribers?

How do you know your answer is reasonable?

HIP-HOP MATH ANSWER GUIDE

Building a Fan Base: One Million Subscribers

Scenario: Kendrick Lamar's social media manager wants to build a newsletter list that exceeds 1 million subscribers. Currently, Kendrick Lamar has 70,000 people on his subscription list.

To hit her goal, the social media manager decides to run a new promotion. All fans who sign up for Kendrick Lamar's online newsletter will receive one free ringtone.

The results of the campaign demonstrate a pattern:

- On day 1, 180,000 new subscribers joined the list.
- On day 2, 172,500 new subscribers joined the list.
- On day 3, 160,000 new subscribers joined the list.
- On day 4, 142,500 new subscribers joined the list.
- On day 5, 120,000 new subscribers joined the list.

If day 1 of the campaign started on a Friday and the pattern of new subscribers to Kendrick Lamar's e-mail list continues at the same rate, on what day of the week will the social media manager reach her goal of exceeding 1 million subscribers?

Step 1: Write down all the numbers that are given. Leave some space between them. Are the numbers increasing or decreasing?

180,000 172,500 160,000 142,500 120,000 (The number of subscribers are decreasing.)

Step 2: Look at the first two numbers. Ask yourself, "How did they get from the first amount to the second?" Write your answer between the two numbers.

To get from 180,000 to 172,500, the amount subtracted is 7,500.

180,000 172,500 160,000 142,500 120,000
-7,500

Step 3: Look at the second and third numbers. Ask yourself, "How did they get from the second amount to the third?" Write your answer between the two numbers.

To get from 172,500 to 160,000, the amount subtracted is 12,500.

180,000 172,500 160,000 142,500 120,000
-7,500 -12,500

Step 4: Compare what you've written. Do you see a pattern forming?

Not really sure yet, so I'll keep going ...

Step 5: One more time—repeat for the third and fourth number (two numbers don't make a pattern; you need at least three).

- Do you see a pattern yet?

To get from 160,000 to 142,500, the amount subtracted is 17,500.

180,000 172,500 160,000 142,500 120,000
-7,500 -12,500 -17,500

Yes, I see that for each new day, the change in the number of new subscribers ***decreases by 5,000.*** *So even though the number of people who joined the list decreases each day, the amount of the change is actually getting larger each day.*

Answer Guide Step 6 continued on next page

Step 6: Use the pattern to answer the question.

In keeping with the pattern:

Day	New Subscribers	Change	Total # of Subscribers
Before Day 1			70,000
Day 1 (Friday)	180,000		250,000
		−7,500	
Day 2 (Saturday)	172,500		422,500
		−12,500	
Day 3 (Sunday)	160,000		582,500
		−17,500	
Day 4 (Monday)	142,500		725,000
		−22,500	
Day 5 (Tuesday)	120,000		845,000
		−27,500	
Day 6 (Wednesday)	92,500		937,500
		−32,500	
Day 7 (Thursday)	60,000		997,500
		−37,500	
Day 8 (Friday)	22,500		1,020,000

So on day 8, Friday, there will be over 1 million subscribers.

Step 7: How do you know your answer is reasonable?

(Answers will vary.) *If I estimate that 100,000 people subscribe each day for 8 days, that's 800,000 subscribers. Add that to the 70,000 that were already on the list, and it's 870,000, which is pretty close to the goal of 1 million subscribers.*

HIP-HOP MATH STUDENT CHALLENGE

1 Million vs. Just 1

PLATINUM PROBLEM

Scenario: Kendrick Lamar's social media manager wants to increase the number of hits that Lamar's website gets in the first month of the album release. She interviews two search engine optimization firms. The first firm claims that its marketing strategy can secure 1 million hits by the end of the month (30 days). The second firm focuses on a slow build, and claims it will get only 1 website hit on the first day, 2 on the second, 4 on the third, 8 on the fourth, and 16 on the fifth. If this pattern continues, which company will produce more website hits for Kendrick Lamar's website by the end of the month?

HIP-HOP MATH ANSWER GUIDE

1 Million vs. Just 1

PLATINUM PROBLEM

Scenario: Kendrick Lamar's social media manager wants to increase the number of hits that Lamar's website gets in the first month of the album release. She interviews two search engine optimization firms. The first firm claims that its marketing strategy can secure 1 million hits by the end of the month (30 days). The second firm focuses on a slow build, and claims it will get only 1 website hit on the first day, 2 on the second, 4 on the third, 8 on the fourth, and 16 on the fifth. If this pattern continues, which company will produce more website hits for Kendrick Lamar's website by the end of the month?

1 Million vs. Just 1

(I see that this is a "Find the pattern" problem and that the number of hits doubles each day.)

Day 1	1
Day 2	2
Day 3	4
Day 4	8
Day 5	16
Day 6	32
Day 7	64
Day 8	128
Day 9	256
Day 10	512
Day 11	1,024
Day 12	2,048
Day 13	4,096
Day 14	8,192
Day 15	16,384
Day 16	32,768
Day 17	65,536
Day 18	131,072
Day 19	262,144
Day 20	524,288
Day 21	1,048,576
Day 22	2,097,152

I can stop after day 21, as the firm with the "slow build" will clearly produce more website hits over 30 days.

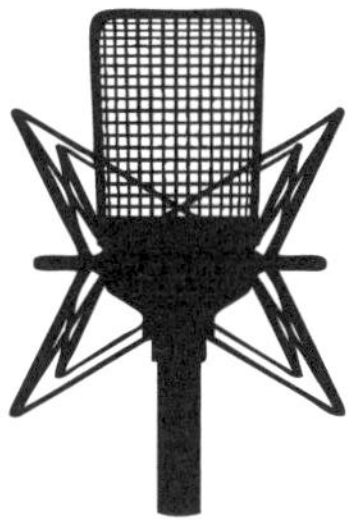

TWELVE TOP TERMS: KEY ACADEMIC VOCABULARY

Anticipate
To expect or predict.

Budget
An estimate of income and expenditure for a set period of time.

Compare
To note the similarity or dissimilarity between two or more things.

Data
Facts and statistics collected together for reference or analysis.

Decrease
To make or become smaller in size, amount, intensity, or degree.

Exceed
To be greater in size or number than.

Growth
The process of increasing in number or size.

Increase
To make or become bigger in size, amount, intensity, or degree.

Pattern
A predictable change.

Predict
To forecast, or to express what will happen in the future.

Repeating
Occurring in the same way.

Surpass
To exceed or be greater than.

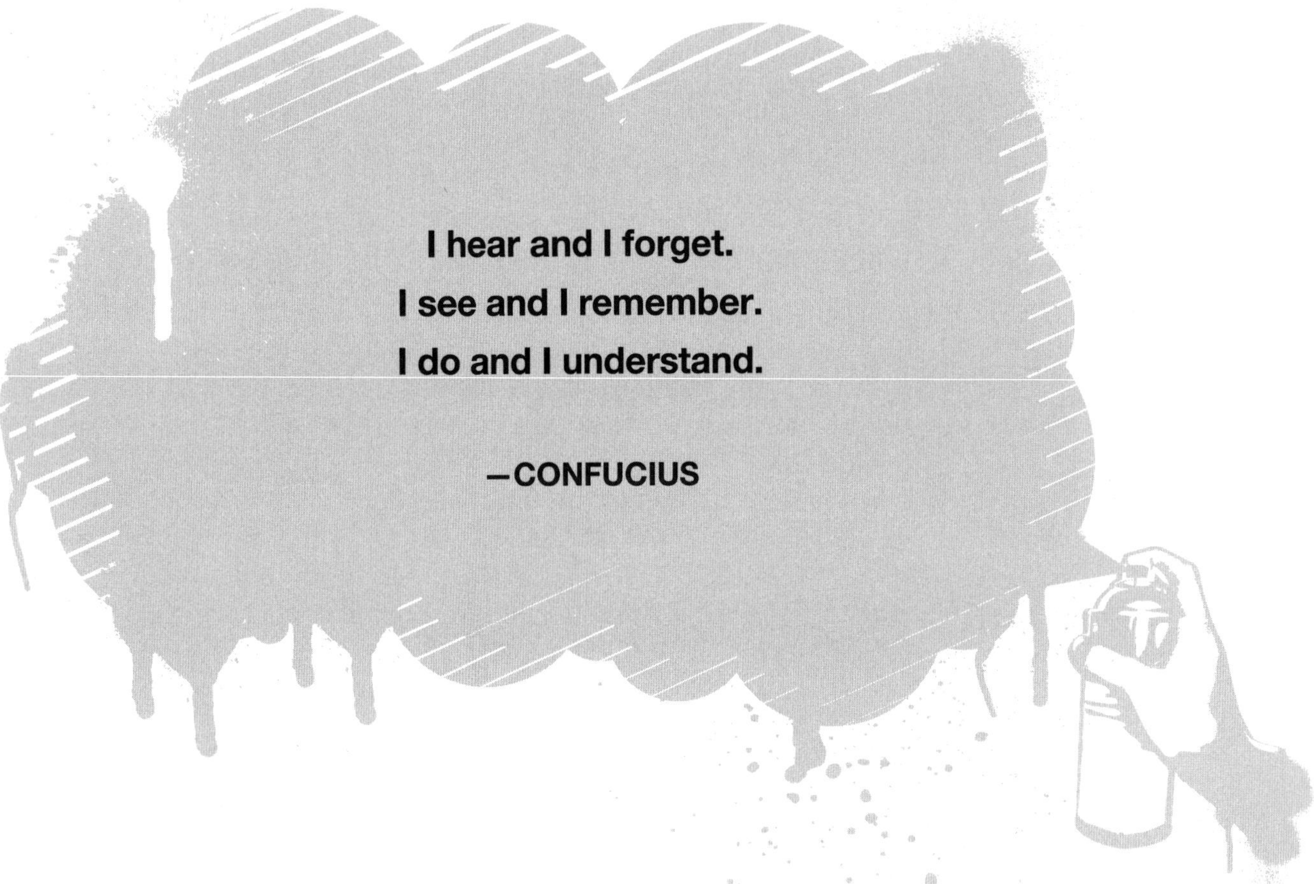
I hear and I forget.
I see and I remember.
I do and I understand.
—CONFUCIUS

UNIT 6

DRAW A PICTURE

Drawing a picture helps you:

organize information ▪ visualize what is happening ▪ "see" your solution

Step 1: Read the entire problem.

Step 2: Underline the key parts.

Step 3: Circle the question or task.

Step 4: Visualize and draw the scenario.

Step 5: Revisit: What's the question?

Step 6: Choose your operation(s) and find the answer.

Step 7: How do you know your answer is reasonable?

HIP-HOP MATH STUDENT CHALLENGE

Showtime with Beyoncé

Scenario: Beyoncé's new tour will see her criss-crossing the country, performing in a series of sold-out venues.

On the first leg of the tour, she will visit four cities in the Northeast: New York, Philadelphia, Washington, D.C., and Boston. But there are multiple schedule considerations that the tour manager must take into account before planning the dates for Beyoncé's shows.

- The New York City show must occur before the Washington, D.C., show.
- The Boston show must occur between the New York City and Washington, D.C., shows.
- The Philadelphia show must occur after the Washington, D.C., show.

What city will be first on Beyoncé's Northeast tour?

Step 1: Read the entire problem.

Step 2: Underline the key parts.

Step 3: Circle the question or task.

Step 4: Visualize and draw the scenario.

Step 5: Revisit: What's the question?

Step 6: Choose your operation(s) and find the answer.

Step 7: How do you know your answer is reasonable?

____

____

HIP-HOP MATH ANSWER GUIDE

Showtime with Beyoncé

Scenario: Beyoncé's new tour will see her criss-crossing the country, performing in a series of sold-out venues.

On the first leg of the tour, she will visit four cities in the Northeast: New York, Philadelphia, Washington, D.C., and Boston. But there are multiple schedule considerations that the tour manager must take into account before planning the dates for Beyoncé's shows.

- The New York City show must occur before the Washington, D.C., show. *Part 1*
- The Boston show must occur between the New York City and Washington, D.C., shows. *Part 2*
- The Philadelphia show must occur after the Washington, D.C., show. *Part 3*

What city will be first on Beyoncé's Northeast tour?

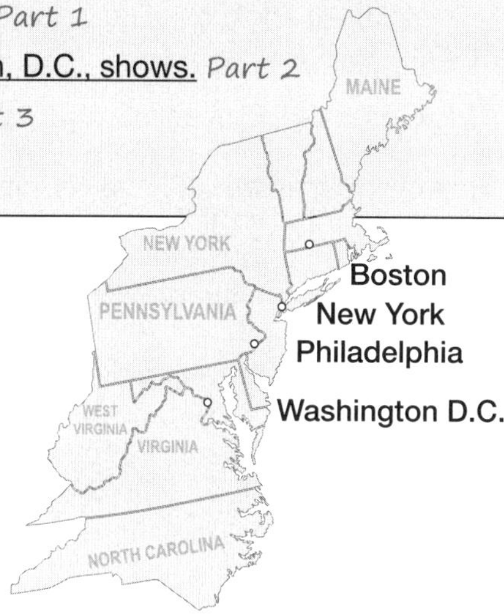

Step 1: Read the entire problem.

Step 2: Underline the key parts.

Step 3: Circle the question or task.

Step 4: Visualize and draw the scenario.

Part 1	*Part 2*	*Part 3*	*Final Result*
(NYC before D.C.)	*(Boston between NYC & D.C.)*	*(Phil after D.C.)*	
	Boston ↓	*Phil* ↘	
NYC D.C.	*NYC D.C.*	*NYC Boston D.C.*	*NYC Boston D.C. Phil*

Step 5: Revisit: What's the question?

What city will be first on Beyoncé's Northeast tour?

Step 6: Choose your operation(s) and find the answer.

New York City will be the first stop on the Northeast tour.

Step 7: How do you know your answer is reasonable?

(Answers will vary.) I went back and made sure the answer fit all of the requirements.

HIP-HOP MATH STUDENT CHALLENGE

Showtime with Beyoncé: Touring the West Coast

Scenario: Beyoncé's new tour will see her criss-crossing the country, performing in a series of sold-out venues.

On the second leg of the tour, she will visit four cities on the West Coast: Los Angeles, San Francisco, Seattle, and Portland. But there are multiple schedule considerations that the tour manager must take into account before planning the dates for Beyoncé's shows:

- The Seattle show must occur after the Los Angeles show.
- The San Francisco show must occur between the Seattle and Los Angeles shows.
- The Portland show must occur between the San Francisco and Seattle shows.

What city will be the last stop on this leg of Beyoncé's West Coast tour?

Step 1: Read the entire problem.

Step 2: Underline the key parts.

Step 3: Circle the question or task.

Step 4: Visualize and draw the scenario.

Step 5: Revisit: What's the question?

Step 6: Choose your operation(s) and find the answer.

Step 7: How do you know your answer is reasonable?

HIP-HOP MATH ANSWER GUIDE

Showtime with Beyoncé: Touring the West Coast

Scenario: Beyoncé's new tour will see her criss-crossing the country, performing in a series of sold-out venues.

On the second leg of the tour, she will visit four cities on the West Coast: Los Angeles, San Francisco, Seattle, and Portland. But there are multiple schedule considerations that the tour manager must take into account before planning the dates for Beyoncé's shows:

- The Seattle show must occur after the Los Angeles show. *Step 1*
- The San Francisco show must occur between the Seattle and Los Angeles shows. *Step 2*
- The Portland show must occur between the San Francisco and Seattle shows. *Step 3*

What city will be the last stop on this leg of Beyoncé's West Coast tour?

Step 1: Read the entire problem.

Step 2: Underline the key parts.

Step 3: Circle the question or task.

Step 4: Visualize and draw the scenario.

Step 1	*Step 2*	*Step 3*	*Final Result*
(Sea after LA)	*(SFO between Sea & LA)*	*(Port between SFO and Sea Port)*	
	SFO ↓	*Port* ↘	
LA Sea	*LA Sea*	*LA SFO Sea*	*LA SFO Port Sea*

Step 5: Revisit: What's the question?
What city will be the last stop on Beyoncé's West Coast tour?

Step 6: Choose your operation(s) and find the answer.
The last stop on the West Coast tour will be Seattle.

Step 7: How do you know your answer is reasonable?
(Answers will vary.) I went back and made sure the answer fit all of the requirements.

HIP-HOP MATH STUDENT CHALLENGE

Showtime with Beyoncé: Touring the South East

Scenario: The third leg of Beyoncé's tour will see her visit five cities in the Southeast: Miami, Atlanta, Tampa, New Orleans, and Fort Lauderdale. There are multiple schedule considerations that the tour manager must take into account before planning the dates for Beyoncé's shows:

- The Ft. Lauderdale show must happen after the Miami show.
- The Atlanta show must be after New Orleans but before Miami.
- The Miami show must be after the Tampa show.
- The Tampa show must be directly after Atlanta.

In what order will the five cities for Beyoncé's tour through the Southeast be scheduled?

Step 1: Read the entire problem.

Step 2: Underline the key parts.

Step 3: Circle the question or task.

Step 4: Visualize and draw the scenario.

Step 5: Revisit: What's the question?

Step 6: Choose your operation(s) and find the answer.

Step 7: How do you know your answer is reasonable?

HIP-HOP MATH ANSWER GUIDE

Showtime with Beyoncé: Touring the South East

Scenario: The third leg of Beyoncé's tour will see her visit five cities in the Southeast: Miami, Atlanta, Tampa, New Orleans, and Fort Lauderdale. There are multiple schedule considerations that the tour manager must take into account before planning the dates for Beyoncé's shows:

- The Ft. Lauderdale show must happen after the Miami show. *Part 1*
- The Atlanta show must be after New Orleans but before Miami. *Part 2*
- The Miami show must be after the Tampa show. *Part 3*
- The Tampa show must be directly after Atlanta.. *Part 4*

In what order will the five cities for Beyoncé's tour through the Southeast be scheduled?

Step 1: Read the entire problem.

Step 2: Underline the key parts.

Step 3: Circle the question or task.

Step 4: Visualize and draw the scenario.

Part 1	Part 2	Part 3	Part 4	Final Result
(Ft after Mi)	(Atl after NO before Mi)	(Tam must be before Mi)	(Tam after Atl)	
	NO Atl	No Atl | Mi Ft	No Atl | Mi Ft	
Mi Ft	Mi Ft	Tam ←	Tam ←	NO Atl Tam Mi Ft

Step 5: Revisit: What's the question?

In what order will the five cities for Beyoncé's tour through the Southeast be scheduled?

Step 6: Choose your operation(s) and find the answer.

The order of the cities for the Southeast tour will be New Orleans, Atlanta, Tampa, Miami, Ft. Lauderdale.

Step 7: How do you know your answer is reasonable?

(Answers will vary.) I went back and made sure the answer fit all of the requirements.

HIP-HOP MATH STUDENT CHALLENGE

Showtime with Beyoncé: Clothing Combinations

Scenario: In between concerts, Beyoncé will often make appearances on talk shows and conduct interviews on entertainment news shows. Her wardrobe stylist has decided that for each interview, Beyoncé will wear pants, a blouse, and a coat. The stylist does not want her to repeat the exact same outfit for any of these appearances, however. Beyoncé has three different colors of pants: red, black, and white. She has four different blouses: blue, green, yellow, and purple. And she has three jackets: solid, striped, and checkered.

How many different combinations of pants, blouse, and jacket can Beyoncé wear without repeating the exact same outfit?

Step 1: Read the entire problem.

Step 2: Underline the key parts.

Step 3: Circle the question or task.

Step 4: Visualize and draw the scenario.

Step 5: Revisit: What's the question?

Step 6: Choose your operation(s) and find the answer.

Step 7: How do you know your answer is reasonable?

HIP-HOP MATH ANSWER GUIDE

Showtime with Beyoncé: Clothing Combinations

Scenario: In between concerts, Beyoncé will often make appearances on talk shows and conduct interviews on entertainment news shows. Her wardrobe stylist has decided that for each interview, Beyoncé will wear pants, a blouse, and a coat. The stylist does not want her to repeat the exact same outfit for any of these appearances, however. Beyoncé has three different colors of pants: red, black, and white. She has four different blouses: blue, green, yellow, and purple. And she has three jackets: solid, striped, and checkered.

How many different combinations of pants, blouse, and jacket can Beyoncé wear without repeating the exact same outfit?

Step 1: Read the entire problem.

Step 2: Underline the key parts.

Step 3: Circle the question or task.

Step 4: Visualize and draw the scenario.

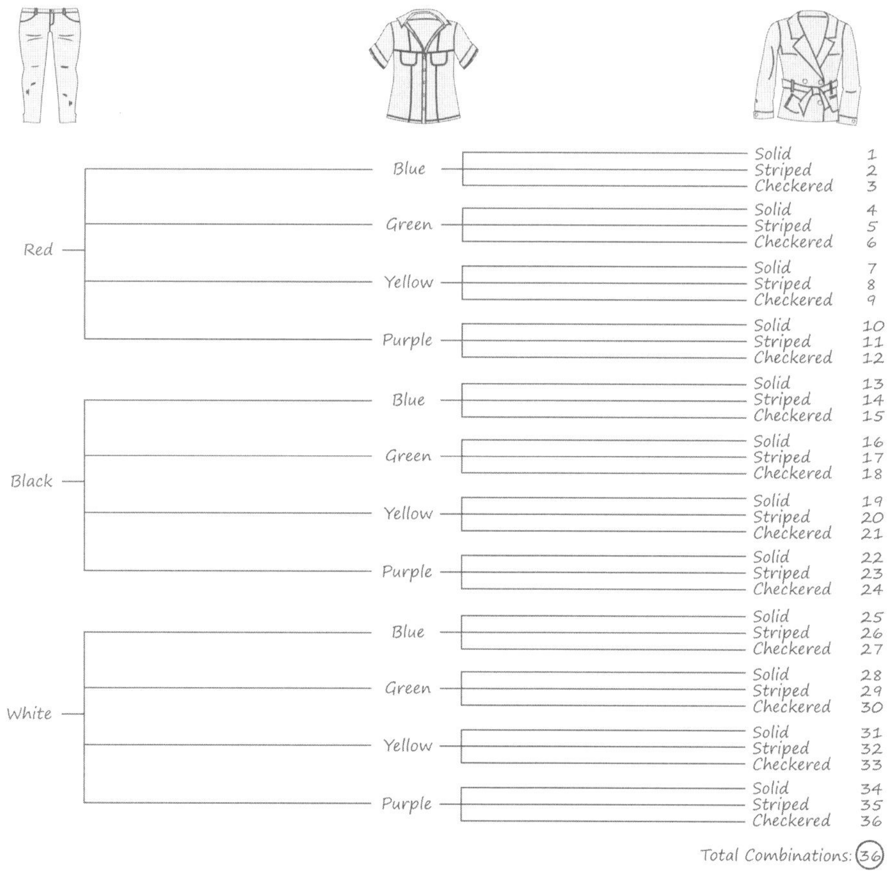

Step 5: Revisit: What's the question?
How many different combinations of pants, blouse, and jacket can Beyoncé wear without repeating the exact same outfit?

Step 6: Choose your operation(s) and find the answer.
When I add all of the different combinations together, I get 36 different possible outfits.

Step 7: How do you know your answer is reasonable?
(Answers will vary.) Because I wrote out all of the possible outfits.

HIP-HOP MATH STUDENT CHALLENGE

Showtime with Beyoncé: Fireworks Regulations

Scenario: Beyoncé is exceptionally popular in Canada, so the tour manager decides to schedule a series of performances for her Canadian fans. However, the Canadian fire marshall has mandated that for Beyoncé's tour there will be strict regulations on the amount of fireworks that can be used for the show's grand finale.

½ of the shows will have no restrictions because they take place in outdoor arenas. ¼ of the shows will allow no fireworks whatsoever because the indoor venues have roofs that are too low. The remaining 3 shows on the tour will allow a small amount of fireworks but not grand explosions that exceed 5 meters (16.4 feet) in height.

How many shows will allow Beyoncé to incorporate at least some amount of fireworks during her performance?

How do you know your answer is reasonable?

HIP-HOP MATH ANSWER GUIDE

Showtime with Beyoncé: Fireworks Regulations

Scenario: Beyoncé is exceptionally popular in Canada, so the tour manager decides to schedule a series of performances for her Canadian fans. However, the Canadian fire marshall has mandated that for Beyoncé's tour there will be strict regulations on the amount of fireworks that can be used for the show's grand finale.

½ of the shows will have no restrictions because they take place in outdoor arenas. ¼ of the shows will allow no fireworks whatsoever because the indoor venues have roofs that are too low. The remaining 3 shows on the tour will allow a small amount of fireworks but not grand explosions that exceed 5 meters (16.4 feet) in height.

How many shows will allow Beyoncé to incorporate at least some amount of fireworks during her performance?

Step 1: Read the entire problem.

Step 2: Underline the key parts.

Step 3: Circle the question or task.

Step 4: Visualize and draw the scenario.

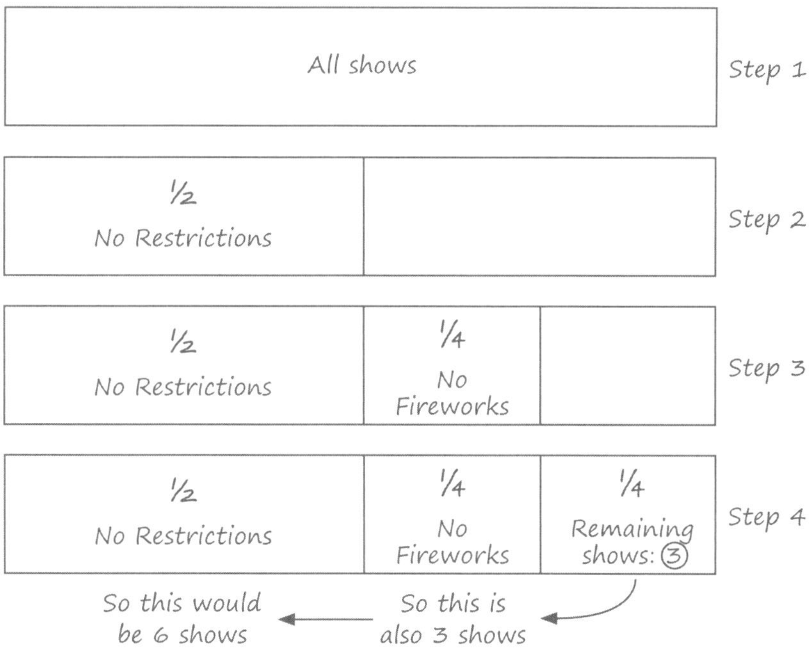

Step 5: Revisit: What's the question?

How many shows will allow Beyoncé to incorporate at least some amount of fireworks?

Step 6: Choose your operation(s) and find the answer.

So, the number of shows that Beyoncé can have at least some fireworks at is ⑨.

Step 7: How do you know your answer is reasonable?

(Answers will vary.) I know it has to be a number bigger than 3, but less than 12 (the total number of shows), and 9 is in between 3 and 12.

HIP-HOP MATH STUDENT CHALLENGE

Showtime with Beyoncé: Usable Stage Space

Scenario: Since Beyoncé's Madison Square Garden concert will be filmed for an upcoming HBO special, there are specific stage parameters within which Beyoncé must stay in order for the HBO cameras to always be able to remain focused on her during the show.

Mike, Beyoncé's lead roadie, decides to mark the points beyond which Beyoncé can not cross with white masking tape. He decides that every three feet he will place a big letter X on the floor so Beyoncé and her dancers can easily see the boundaries.

The usable stage space for the HBO concert special has a length of 21 feet and a width of 18 feet. If Mike puts a big letter X on the stage's floor every 3 feet along the perimeter of the boundaries, how many total big X marks will Mike need to make?

How do you know your answer is reasonable?

____

____

____

HIP-HOP MATH ANSWER GUIDE

Showtime with Beyoncé: Usable Stage Space

Scenario: Since Beyoncé's Madison Square Garden concert will be filmed for an upcoming HBO special, there are specific stage parameters within which Beyoncé must stay in order for the HBO cameras to always be able to remain focused on her during the show.

Mike, Beyoncé's lead roadie, decides to mark the points beyond which Beyoncé can not cross with white masking tape. He decides that every three feet he will place a big letter X on the floor so Beyoncé and her dancers can easily see the boundaries.

The usable stage space for the HBO concert special has a length of 21 feet and a width of 18 feet. If Mike puts a big letter X on the stage's floor every 3 feet along the perimeter of the boundaries, how many total big X marks will Mike need to make?

Step 1: Read the entire problem.

Step 2: Underline the key parts.

Step 3: Circle the question or task.

Step 4: Visualize and draw the scenario.

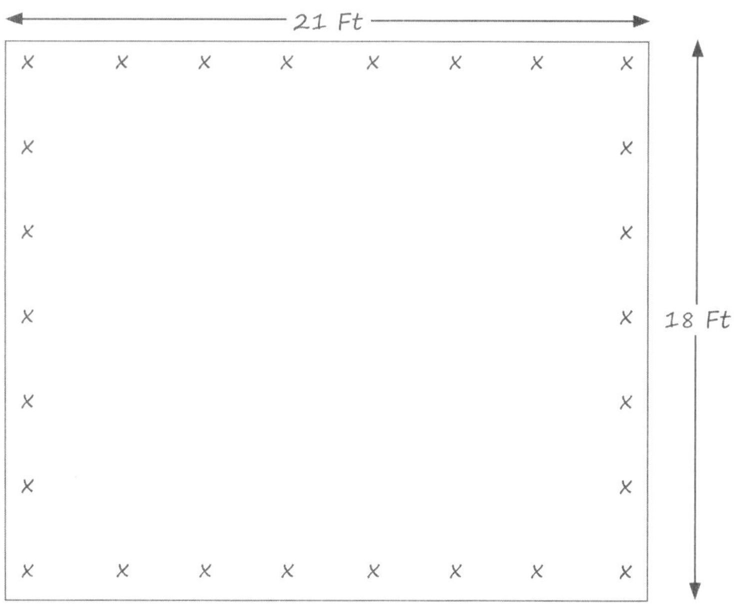

Step 5: Revisit: What's the question?

How many total X marks will Mike need to make?

Step 6: Choose your operation(s) and find the answer.

I counted 26 white Xs in total.

Step 7: How do you know your answer is reasonable?

(Answers will vary.) *The entire perimeter of the boundary is 78 feet, and there is an "X" every 3 feets. Since 78 ÷ 3 is 26, I know this is a reasonable answer.*

HIP-HOP MATH STUDENT CHALLENGE

Showtime with Beyoncé: VIP Backstage Party

Scenario: After the HBO concert special at Madison Square Garden, Beyoncé's record label is going to host a VIP backstage after-party for 100 people.

The caterers will bring in 6 round tables, 4 square tables, 2 long picnic style tables and 8 sofas.

- 6 people can sit at each round table.
- 4 people can sit at each square table.
- Two identical picnic style tables can be placed end-to-end so that 24 total people can be seated on their benches.

If all the guests are seated, and all of the tables are filled to capacity, how many people will sit on each sofa if all 8 of the sofas accommodate the exact same number of guests?

How do you know your answer is reasonable?

___

___

___

HIP-HOP MATH ANSWER GUIDE

Showtime with Beyoncé: VIP Backstage Party

Scenario: After the HBO concert special at Madison Square Garden, Beyoncé's record label is going to host a VIP backstage after-party for 100 people.

The caterers will bring in 6 round tables, 4 square tables, 2 long picnic style tables and 8 sofas.

- 6 people can sit at each round table.
- 4 people can sit at each square table.
- Two identical picnic style tables can be placed end-to-end so that 24 total people can be seated on their benches.

If all the guests are seated, and all of the tables are filled to capacity, how many people will sit on each sofa if all 8 of the sofas accommodate the exact same number of guests?

Step 1: Read the entire problem.

Step 2: Underline the key parts.

Step 3: Circle the question or task.

Step 4: Visualize and draw the scenario.

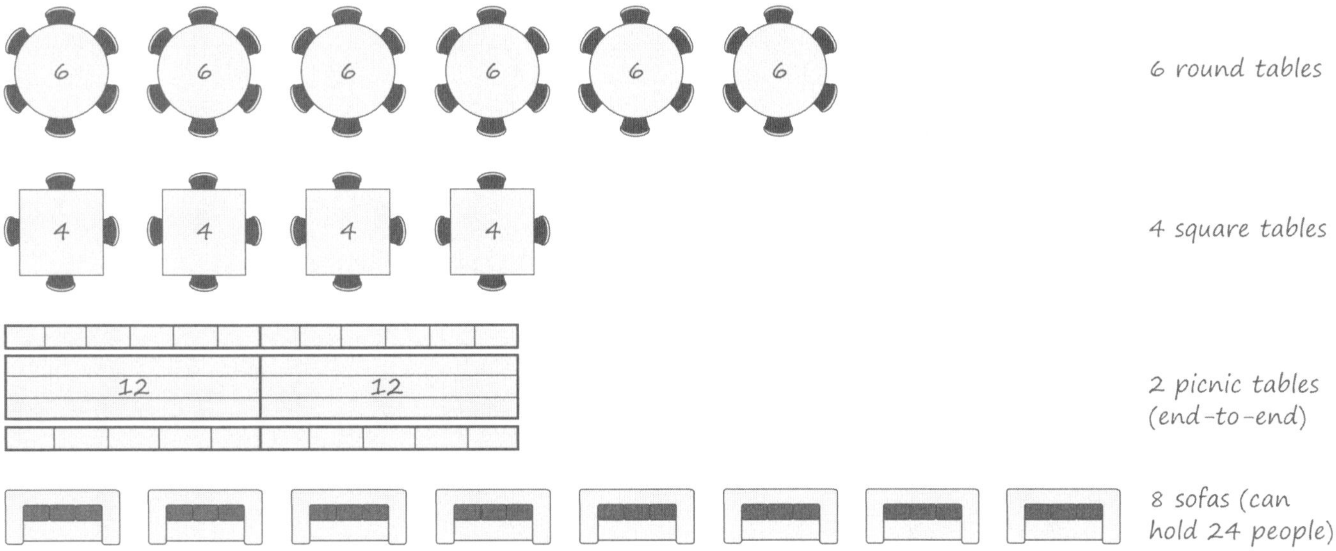

Step 5: Revisit: What's the question?
How many people can sit on each sofa (all sofas hold the same # of people)

Step 6: Choose your operation(s) and find the answer.
6 rounds hold 6 people each: 6 x 6 = 36
4 squares hold 4 people each: 4 x 4 = 16
2 picnics hold 24 total
That's 36 + 16 + 24 = 76 seats
There will be 100 people, so 100-76 = 24 sofa seats.
There are 8 sofas, so 24 ÷ 8 = 3 seats per sofa

Step 7: How do you know your answer is reasonable?
(Answers will vary.) *My sofa at home seats 3 people, so this sounds reasonable.*

HIP-HOP MATH STUDENT CHALLENGE

Ticket Sales!

PLATINUM PROBLEM

Scenario: For the final stop on Beyoncé's tour, she is going to perform in a small venue.

The theater has a maximum seating capacity of 1,500 people and the show is expected to sell out, but the challenge for the theater owner is to price all the seats so that at least $300,000 in revenue is generated. There are 5 categories of seats available.

- Orchestra Pit: 150 seats (closest to the Stage)
- Section A: seats 400 people (directly behind the Orchestra Pit)
- Section B: seats 400 people (directly behind Section A)
- Section C: seats 400 people (directly behind the Section B)
- General Seating Benches: 150 seats (The general seating benches are furthest from the stage)

Your task is to assign a price to all 1,500 seats while staying within the following parameters:

- The closer to the stage, the higher the ticket price.
- No two sections can have the same ticket price.
- The total revenue from the ticket sales must be between $300,000 and $350,000.

HIP-HOP MATH ANSWER GUIDE

Ticket Sales!

PLATINUM PROBLEM

Scenario: For the final stop on Beyoncé's tour, she is going to perform in a small venue.

The theater has a maximum seating capacity of 1,500 people and the show is expected to sell out, but the challenge for the theater owner is to price all the seats so that at least $300,000 in revenue is generated. There are 5 categories of seats available.

- Orchestra Pit: 150 seats (closest to the Stage)
- Section A: seats 400 people (directly behind the Orchestra Pit)
- Section B: seats 400 people (directly behind Section A)
- Section C: seats 400 people (directly behind the Section B)
- General Seating Benches: 150 seats (The general seating benches are furthest from the stage)

Your task is to assign a price to all 1,500 seats while staying within the following parameters:

- The closer to the stage, the higher the ticket price.
- No two sections can have the same ticket price.
- The total revenue from the ticket sales must be between $300,000 and $350,000.

(Answers will vary.)

STAGE

Orchestra 150 seats x $400 per seat	= $60,000
Section A 400 seats x $300 per seat	= $120,000
Section B 400 seats x $200 per seat	= $80,000
Section C 400 seats x $100 per seat	= $40,000
General Seating 150 seats x $50 per seat	= $7,500
Total:	= $307,500

TWELVE TOP TERMS: KEY ACADEMIC VOCABULARY

Accommodate
To provide enough space for.

Assign
To designate something for a specific purpose.

Boundary
A line that marks the limits of an area.

Category
A group of people or things that have similar characteristics.

Length
The measure of something from end to end.

Maximum
The greatest or highest amount possible.

Parameters
Factors that can be measured or made clear.

Perimeter
The distance around an object.

Remaining
The amount or number of items that are unused or still left.

Revenue
Money taken in by a store or service.

Vary
To differ in size and amount.

Width
The measure of something from side to side.

UNIT 7

MAKE AN ORGANIZED LIST

Making an organized list helps you:

arrange information systematically ▪ see different combinations ▪ keep an accurate record

Step 1: Identify all of your beginning elements.

Step 2: Select the first element and list every possible combination you can create with it.

Step 3: Select a second element and list every possible combination you can create with it.

**Step 4: Select a third element and list every possible combination you can create with it.
Keep a lookout for combinations you have already made.**

Step 5: If there are more than 3 elements, repeat the process until all elements have been used.

Step 6: Make sure you have eliminated duplicates before submitting a final answer.

Step 7: How do you know your answer is reasonable?

HIP-HOP MATH STUDENT CHALLENGE
Beat-Making

Scenario: As an up-and-coming beat maker, DJ TripppleBeatz has been given a chance to submit his work to Kanye West for a forthcoming album. The guidelines are very specific:

The DJ must select a combination that contains one of the following 3 tempos as matched to one of the following 4 moods:

Tempo options:

- slow
- medium
- fast

Mood options:

- soulful
- upbeat
- dark
- jazzy

How many total possible combinations of beats is it possible for DJ TripppleBeatz to create that follow these specific guidelines?

Step 1: Identify all of your beginning elements.

Step 2: Select the first element and list every possible combination you can create with it.

Step 3: Select a second element and list every possible combination you can create with it.

Step 4: Select a third element and list every possible combination you can create with it. Keep a lookout for combinations you have already made.

Step 5: If there are more than 3 elements, repeat the process until all elements have been used.

Step 6: Make sure you have eliminated duplicates before submitting a final answer.

Step 7: How do you know your answer is reasonable?

____

____

____

HIP-HOP MATH ANSWER GUIDE

Beat-Making

Scenario: As an up-and-coming beat maker, DJ TripppleBeatz has been given a chance to submit his work to Kanye West for a forthcoming album. The guidelines are very specific:

The DJ must select a combination that contains one of the following 3 tempos as matched to one of the following 4 moods:

Tempo options:

- slow
- medium
- fast

Mood options:

- soulful
- upbeat
- dark
- jazzy

How many total possible combinations of beats is it possible for DJ TripppleBeatz to create that follow these specific guidelines?

Step 1: Identify all of your beginning elements.

The beginning elements are the tempo options:

slow (element #1) *medium (element #2)* *fast (element #3)*

Step 2: Select the first element and list every possible combination you can create with it.

I'll start with the first tempo option, ***slow****, and combine it with each mood option:*

slow *– soulful*
slow *– upbeat*
slow *– dark*
slow *– jazzy*
(That's 4 combinations so far.)

Step 3: Select a second element and list every possible combination you can create with it.

Next, I'll select the second tempo option, ***medium****, and combine it with each mood option:*

slow – soulful
slow – upbeat
slow – dark
slow – jazzy
(That's 4 combinations so far.)

medium *– soulful*
medium *– upbeat*
medium *– dark*
medium *– jazzy*
(That's 4 more combinations, so 8 total so far.)

Answer Guide Step 4 continued on next page

Step 4: Select a third element and list every possible combination you can create with it. Keep a lookout for combinations you have already made.

Next, I'll select the third tempo option, ***fast****, and combine it with each mood option:*

slow – soulful
slow – upbeat
slow – dark
slow – jazzy
(That's 4 combinations so far.)

medium – soulful
medium – upbeat
medium – dark
medium – jazzy
(That's 4 more combinations, so 8 total so far.)

fast *– soulful*
fast *– upbeat*
fast *– dark*
fast *– jazzy*
(That's 4 more combinations, so 12 total combinations.)

Step 5: If there are more than 3 elements, repeat the process until all elements have been used.

There are only 3 elements.

Step 6: Make sure you have eliminated duplicates before submitting a final answer.

slow – soulful
slow – upbeat
slow – dark
slow – jazzy
(That's 4 combinations so far.)

medium – soulful
medium – upbeat
medium – dark
medium – jazzy
(That's 4 more combinations, so 8 total so far.)

fast – soulful
fast – upbeat
fast – dark
fast – jazzy
(That's ~~4 more combi~~nations, so 12 total combinations.)

There are no duplicates, so 12 total combinations.

Step 7: How do you know your answer is reasonable?

(Answers will vary.) *There are 3 different tempos and 4 different moods; since 3 x 4 = 12 and there are no combinations that repeat, this answer is reasonable.*

HIP-HOP MATH STUDENT CHALLENGE

Beat-Making: Combination and Samples

Scenario: As an up-and-coming beat maker, DJ TripppleBeatz has made it to the second round of auditions and has been given a chance to submit his work to Kanye West for an upcoming album. The guidelines are very specific:

The DJ must select a combination that contains one of the following 5 musical samples as matched to one of the following 3 musical instruments:

Music samples:

- rock
- gospel
- reggae
- disco
- blues

Instrument options:

- drums
- guitar
- synthesizer

How many total combinations is it possible for DJ TripppleBeatz to create that follow these specific guidelines?

Step 1: Identify all of your beginning elements.

Step 2: Select the first element and list every possible combination you can create with it.

Step 3: Select a second element and list every possible combination you can create with it.

Step 4: Select a third element and list every possible combination you can create with it. Keep a lookout for combinations you have already made.

Step 5: If there are more than 3 elements, repeat the process until all elements have been used.

Step 6: Make sure you have eliminated duplicates before submitting a final answer.

Step 7: How do you know your answer is reasonable?

____

____

____

HIP-HOP MATH ANSWER GUIDE

Beat-Making: Combination and Samples

Scenario: As an up-and-coming beat maker, DJ TripppleBeatz has made it to the second round of auditions and has been given a chance to submit his work to Kanye West for an upcoming album. The guidelines are very specific:

The DJ must select a combination that contains one of the following 5 musical samples as matched to one of the following 3 musical instruments:

Music samples:

- rock
- gospel
- reggae
- disco
- blues

Instrument options:

- drums
- guitar
- synthesizer

How many total combinations is it possible for DJ TripppleBeatz to create that follow these specific guidelines?

Step 1: Identify all of your beginning elements.

The beginning elements are the music samples:
rock (element #1) gospel (element #2) reggae (element #3) disco (element #4) blues (element #5)

Step 2: Select the first element and list every possible combination you can create with it.

*I'll start with the first music sample, **rock**, and combine it with each instrument option:*

***rock** – drums*
***rock** – guitar*
***rock** – synthesizer*
(That's 3 combinations so far.)

Step 3: Select a second element and list every possible combination you can create with it.

*I'll start with the second music sample, **gospel**, and combine it with each instrument option:*

rock – drums	***gospel** – drums*
rock – guitar	***gospel** – guitar*
rock – synthesizer	***gospel** – synthesizer*
(That's 3 combinations so far.)	*(That's 3 more combinations, so 6 total so far.)*

Step 4: Select a third element and list every possible combination you can create with it. Keep a lookout for combinations you have already made.

*I'll start with the third music sample, **reggae**, and combine it with each instrument option:*

rock – drums	*gospel – drums*	***reggae** – drums*
rock – guitar	*gospel – guitar*	***reggae** – guitar*
rock – synthesizer	*gospel – synthesizer*	***reggae** – synthesizer*
(That's 3 combinations so far.)	*(That's 3 more combinations, so 6 total so far.)*	*(That's 3 more combinations, so 9 total so far.)*

Answer Guide Step 4 continued on next page

Step 5: If there are more than 3 elements, repeat the process until all elements have been used.

*The next element is **disco**:*

rock – drums	gospel – drums	reggae – drums	**disco** – drums
rock – guitar	gospel – guitar	reggae – guitar	**disco** – guitar
rock – synthesizer	gospel – synthesizer	reggae – synthesizer	**disco** – synthesizer
(That's 3 combinations so far.)	(That's 3 more combinations, so 6 total so far.)	(That's 3 more combinations, so 9 total so far.)	(That's 3 more combinations, so 12 total so far.)

*The next element is **blues**:*

rock – drums	gospel – drums	reggae – drums	disco – drums	**blues** – drums
rock – guitar	gospel – guitar	reggae – guitar	disco – guitar	**blues** – guitar
rock – synthesizer	gospel – synthesizer	reggae – synthesizer	disco – synthesizer	**blues** – synthesizer
(That's 3 combinations so far.)	(That's 3 more combinations, so 6 total so far.)	(That's 3 more combinations, so 9 total so far.)	(That's 3 more combinations, so 12 total so far.)	(That's 3 more combinations, so 15 total combinations).

Step 6: Make sure you have eliminated duplicates before submitting a final answer.

There are no duplicates, so 15 total combinations.

Step 7: How do you know your answer is reasonable?

(Answers will vary.) *There are 5 different music samples and 3 different instrumental options; since 5 x 3 = 15 and there are no combinations that repeat, this answer is reasonable.*

HIP-HOP MATH STUDENT CHALLENGE

Beat-Making: Categories and Tempos

Scenario: DJ TripppleBeatz suddenly receives a text message that Kanye West has narrowed the number of sample options required (see below) but is also insisting that the tempo of the beat be clearly identified. Here are the new categories:

Tempo options:

- slow
- fast

Music samples:

- gospel
- reggae
- blues

Instrument options:

- drums
- guitar

How many total combinations is it possible for DJ TripppleBeatz to create that follow these specific guidelines?

Step 1: Identify all of your beginning elements.

Step 2: Select the first element and list every possible combination you can create with it.

Step 3: Select a second element and list every possible combination you can create with it.

Step 4: Select a third element and list every possible combination you can create with it. Keep a lookout for combinations you have already made.

Step 5: If there are more than 3 elements, repeat the process until all elements have been used.

Step 6: Make sure you have eliminated duplicates before submitting a final answer.

Step 7: How do you know your answer is reasonable?

HIP-HOP MATH ANSWER GUIDE

Beat-Making: Categories and Tempos

Scenario: DJ TrippleBeatz suddenly receives a text message that Kanye West has narrowed the number of sample options required (see below) but is also insisting that the tempo of the beat be clearly identified. Here are the new categories:

Tempo options:

- slow
- fast

Music samples:

- gospel
- reggae
- blues

Instrument options:

- drums
- guitar

How many total combinations is it possible for DJ TrippleBeatz to create that follow these specific guidelines?

Step 1: Identify all of your beginning elements.

The beginning elements are the tempo options: slow (element #1) and fast (element #2).

Step 2: Select the first element and list every possible combination you can create with it.

*I'll start with the first tempo option, **slow**, and combine it with each music sample and then each instrument option:*

slow – gospel – drums | *slow – reggae – drums* | *slow – blues – drums*

slow – gospel – guitar | *slow – reggae – guitar* | *slow – blues – guitar*

(That's 6 combinations so far.)

Step 3: Select a second element and list every possible combination you can create with it.

*Next, I'll start with the second tempo option, **fast**, and combine it with each music sample and then each instrument option:*

fast – gospel – drums | *fast – reggae – drums* | *fast – blues – drums*

fast – gospel – guitar | *fast – reggae – guitar* | *fast – blues – guitar*

(That's 6 more combinations, so there are 12 total combinations.)

Step 4: Select a third element and list every possible combination you can create with it. Keep a lookout for combinations you have already made.

There is no third element.

Step 5: If there are more than 3 elements, repeat the process until all elements have been used.

Step 6: Make sure you have eliminated duplicates before submitting a final answer.

There are no duplicates., so there are 12 total combinations.

Step 7: How do you know your answer is reasonable?

***(Answers will vary.)** There are 2 different tempo options, 3 different music samples, and 2 different instrument options. Since 2 x 3 x 2 = 12 and there are no combinations that repeat, this answer is reasonable.*

HIP-HOP MATH STUDENT CHALLENGE

Beat-Making: Ordering the Selections

Scenario: DJs are allowed to submit a maximum of three original beats to Kanye West for consideration. After weighing his options, DJ TripppleBeatz has selected the following three samples:

- Sample A (a combination of reggae/drums/slow)
- Sample B (a combination of blues/drums/fast)
- Sample C (a combination of gospel/guitar/slow)

The next challenge for DJ TripppleBeatz is selecting the order in which he wants Kanye West to hear his samples. Choosing which beat Kanye West hears first, second, and third is very important. In how many different orders can DJ TripppleBeatz combine Samples A, B, and C?

Step 1: Identify all of your beginning elements.

Step 2: Select the first element and list every possible combination you can create with it.

Step 3: Select a second element and list every possible combination you can create with it.

Step 4: Select a third element and list every possible combination you can create with it. Keep a lookout for combinations you have already made.

Step 5: If there are more than 3 elements, repeat the process until all elements have been used.

Step 6: Make sure you have eliminated duplicates before submitting a final answer.

Step 7: How do you know your answer is reasonable?

____

____

____

HIP-HOP MATH ANSWER GUIDE

Beat-Making: Ordering the Selections

Scenario: DJs are allowed to submit a maximum of three original beats to Kanye West for consideration. After weighing his options, DJ TripppleBeatz has selected the following three samples:

- Sample A (a combination of reggae/drums/slow)
- Sample B (a combination of blues/drums/fast)
- Sample C (a combination of gospel/guitar/slow)

The next challenge for DJ TripppleBeatz is selecting the order in which he wants Kanye West to hear his samples. Choosing which beat Kanye West hears first, second, and third is very important. In how many different orders can DJ TripppleBeatz combine Samples A, B, and C?

Step 1: Identify all of your beginning elements.

The beginning elements are the three samples:

A (element #1) *B (element #2)* *C (element #3)*

Step 2: Select the first element and list every possible combination you can create with it.

*I'll start with sample **A**:*

The order could be:

***A** →B, C*

→C, B

(That's 2 combinations so far.)

Step 3: Select a second element and list every possible combination you can create with it.

*Next, I'll start with sample **B**:*

The order could be:

A →B, C

→C, B

(That's 2 combinations so far.)

The order could be:

***B** →A, C*

→C, A

(That's 2 more,
so 4 combinations total so far.)

Step 4: Select a third element and list every possible combination you can create with it. Keep a lookout for combinations you have already made.

*Next, I'll start with Sample **C**:*

The order could be:

A →B, C

→C, B

(That's 2 combinations so far.)

The order could be:

B →A, C

→C, A

(That's 2 more,
so 4 combinations total so far.)

The order could be:

***C** →A, B*

→B, A

(That's 2 more,
so 6 total ways of ordering the beats.)

Step 5: If there are more than 3 elements, repeat the process until all elements have been used.

Step 6: Make sure you have eliminated duplicates before submitting a final answer.

There are no duplicates so there are 6 total ways of ordering the beats.

Step 7: How do you know your answer is reasonable?

(Answers will vary.) *There are 3 different beats and 2 different ways each beat can be ordered. Since 3 x 2 = 6 and there are no combinations that repeat, this answer is reasonable.*

HIP-HOP MATH STUDENT CHALLENGE

Beat-Making: Assigning Scores

Scenario: Kanye West decides to evaluate each of the three beats submitted to him by assigning it one of three possible scores:

- 1 point
- 3 points
- 5 points

After all three beats are evaluated, how many different score combinations are possible?

How do you know your answer is reasonable?

HIP-HOP MATH ANSWER GUIDE

Beat-Making: Assigning Scores

Scenario: Kanye West decides to evaluate each of the three beats submitted to him by assigning it one of three possible scores:

- 1 point
- 3 points
- 5 points

After all three beats are evaluated, how many different score combinations are possible?

Step 1: Identify all of your beginning elements.

The beginning elements are the three point scores:

1 point (element #1)   *3 points (element #2)*   *5 points (element #3)*

Step 2: Select the first element and list every possible combination you can create with it.

I'll start with the possible point total combinations when the first beat receives 1 point:

The order could be:

1, 1, 1   1, 3, 1   1, 5, 1
1, 1, 3   1, 3, 3   1, 5, 3
1, 1, 5   1, 3, 5   1, 5, 5

(That's 9 possible combinations so far.)

Step 3: Select a second element and list every possible combination you can create with it.

Next, I'll start with the possible point total combinations when the first beat receives 3 points:

The order could be:

1, 1, 1   1, 3, 1   1, 5, 1
1, 1, 3   1, 3, 3   1, 5, 3
1, 1, 5   1, 3, 5   1, 5, 5

(That's 9 possible combinations so far.)

The order could be:

3, 1, 1   3, 3, 1   3, 5, 1
3, 1, 3   3, 3, 3   3, 5, 3
3, 1, 5   3, 3, 5   3, 5, 5

(That's 9 more possible combinations, for a total of 18 so far.)

Step 4: Select a third element and list every possible combination you can create with it. Keep a lookout for combinations you have already made.

Now I'll start with the possible point total combinations when the first beat receives 5 points:

The order could be:

1, 1, 1   1, 3, 1   1, 5, 1
1, 1, 3   1, 3, 3   1, 5, 3
1, 1, 5   1, 3, 5   1, 5, 5

(That's 9 possible combinations so far.)

The order could be:

3, 1, 1   3, 3, 1   3, 3, 1
3, 1, 3   3, 3, 3   3, 5, 3
3, 1, 5   3, 3, 5   3, 5, 5

(That's 9 more possible combinations, for a total of 18 so far.)

The order could be:

5, 1, 1   5, 3, 1   5, 5, 1
5, 1, 3   5, 3, 3   5, 5, 3
5, 1, 5   5, 3, 5   5, 5, 5

(That's 9 more possible combinations, for a total of 27 possible score combinations.)

Answer Guide Step 5 continued on next page

Step 5: If there are more than 3 elements, repeat the process until all elements have been used.

Step 6: Make sure you have eliminated duplicates before submitting a final answer.

The order could be:

1, 1, 1	1, 3, 1	1, 5, 1
1, 1, 3	1, 3, 3	1, 5, 3
1, 1, 5	1, 3, 5	1, 5, 5

(That's 9 possible combinations so far.)

The order could be:

3, 1, 1	3, 3, 1	3, 3, 1
3, 1, 3	3, 3, 3	3, 5, 3
3, 1, 5	3, 3, 5	3, 5, 5

(That's 9 more possible combinations, for a total of 18 so far.)

The order could be:

5, 1, 1	5, 3, 1	5, 5, 1
5, 1, 3	5, 3, 3	5, 5, 3
5, 1, 5	5, 3, 5	5, 5, 5

(That's 9 more possible combinations, for a total of 27 possible score combinations.)

There are no duplicates, so there are a total of 27 possible score combinations.

Step 7: How do you know your answer is reasonable?

(Answers will vary.) There are 3 different ways the first beat can be scored, 3 different ways the second beat can be scored, and 3 different ways the third beat can be scored. Since 3 x 3 x 3 = 27 and there are no combinations that repeat, this answer is reasonable.

HIP-HOP MATH STUDENT CHALLENGE

Beat-Making: Totaling Point Possibilities

Scenario: As DJ TrippleBeatz knows, Kanye West will be evaluating each of the three beats he submitted by assigning it one of the following possible scores:

- 1 point
- 3 points
- 5 points

After determining each beat's individual score, an overall point total will be determined by adding the three individual beat scores together.

How many different overall point totals is it possible for DJ TripppleBeatz to receive?

How do you know your answer is reasonable?

HIP-HOP MATH ANSWER GUIDE

Beat-Making: Totaling Point Possibilities

Scenario: As DJ TrippleBeatz knows, Kanye West will be evaluating each of the three beats he submitted by assigning it one of the following possible scores:

- 1 point
- 3 points
- 5 points

Close Reading Warning: The previous question on page 137 asks a different question than the one you are seeing here!

After determining each beat's individual score, an overall point total will be determined by adding the three individual beat scores together.

How many different overall point totals is it possible for DJ TripppleBeatz to receive?

Step 1: Identify all of your beginning elements.

The beginning elements are the three point scores:
1 point (element #1) 3 points (element #2) and 5 points (element #3)

Step 2: Select the first element and list every possible combination you can create with it.

*I'll start with the possible point total combinations when the first beat receives **1** point and then add them to get a total. I'll wait until the end to cross out any duplicates.*

1, 1, 1 = 3	1, 3, 1 = 5	1, 5, 1 = 7
1, 1, 3 = 5	1, 3, 3 = 7	1, 5, 3 = 9
1, 1, 5 = 7	1, 3, 5 = 9	1, 5, 5 = 11

Step 3: Select a second element and list every possible combination you can create with it.

*Next, I'll start with the possible point total combinations when the first beat receives **3** points, and then add them to get a total. I'll wait until the end to cross out any duplicates.*

1, 1, 1 = 3	1, 3, 1 = 5	1, 5, 1 = 7	3, 1, 1 = 5	3, 3, 1 = 7	3, 5, 1 = 7
1, 1, 3 = 5	1, 3, 3 = 7	1, 5, 3 = 9	3, 1, 3 = 7	3, 3, 3 = 9	3, 5, 3 = 11
1, 1, 5 = 7	1, 3, 5 = 9	1, 5, 5 = 11	3, 1, 5 = 9	3, 3, 5= 11	3, 5, 5 = 13

Step 4: Select a third element and list every possible combination you can create with it. Keep a lookout for combinations you have already made.

*Next, I'll start with the possible point total combinations when the first beat receives **5** points, and then add them to get a total. I'll wait until the end to cross out any duplicates.*

1, 1, 1 = 3	1, 3, 1 = 5	1, 5, 1 = 7	3, 1, 1 = 5	3, 3, 1 = 7	3, 5, 1 = 7	5, 1, 1 = 7	5, 3, 1 = 9	5, 5, 1 = 11
1, 1, 3 = 5	1, 3, 3 = 7	1, 5, 3 = 9	3, 1, 3 = 7	3, 3, 3 = 9	3, 5, 3 = 11	5, 1, 3 = 9	5, 3, 3 = 11	5, 5, 3 = 13
1, 1, 5 = 7	1, 3, 5 = 9	1, 5, 5 = 11	3, 1, 5 = 9	3, 3, 5= 11	3, 5, 5 = 13	5, 1, 5 = 11	5, 3, 5 = 13	5, 5, 5 = 15

Answer Guide Step 5 continued on next page

Step 5: If there are more than 3 elements, repeat the process until all elements have been used.

Step 6: Make sure you have eliminated duplicates before submitting a final answer.

1, 1, 1 = 3	1, 3, 1 = 5	1, 5, 1 = 7	~~3, 1, 1 = 5~~	~~3, 3, 1 = 7~~	~~3, 5, 1 = 7~~	~~5, 1, 1 = 7~~	~~5, 3, 1 = 9~~	~~5, 5, 1 = 11~~
~~1, 1, 3 = 5~~	~~1, 3, 3 = 7~~	1, 5, 3 = 9	~~3, 1, 3 = 7~~	~~3, 3, 3 = 9~~	~~3, 5, 3 = 11~~	~~5, 1, 3 = 9~~	~~5, 3, 3 = 11~~	~~5, 5, 3 = 13~~
~~1, 1, 5 = 7~~	~~1, 3, 5 = 9~~	1, 5, 5 = 11	~~3, 1, 5 = 9~~	~~3, 3, 5 = 11~~	3, 5, 5 = 13	~~5, 1, 5 = 11~~	~~5, 3, 5 = 13~~	5, 5, 5 = 15

There were quite a few duplicates.

So all of the possibilities for point totals are 3, 5, 7, 9, 11, 13 and 15. That's a total of 7 different possible point totals.

Step 7: How do you know your answer is reasonable?

(Answers will vary.) There are 27 different possible combinations of scores, so I knew the answer had to be less than 27, especially once I eliminated all of the duplicates. So 7 is a reasonable amount.

HIP-HOP MATH STUDENT CHALLENGE

Beat-Making: Ordering Rotations

Scenario: DJ TrippleBeatz has been informed that his work will be included on 25% of Kanye's new 20-track album. Additionally, Kanye, will create a Spotify stream that exclusively features DJ TrippleBeatz's artistry.

If the first song that plays is always the same, how many different ways can Kanye order the rest of the DJ TrippleBeatz Spotify playlist?

How do you know your answer is reasonable?

HIP-HOP MATH ANSWER GUIDE

Beat-Making: Ordering Rotations

Scenario: DJ TripppleBeatz has been informed that his work will be included on 25% of Kanye's new 20-track album. Additionally, Kanye, will create a Spotify stream that exclusively features DJ TripppleBeatz's artistry.

If the first song that plays is always the same, how many different ways can Kanye order the rest of the DJ TripppleBeatz Spotify playlist?

Step 1: Identify all of your beginning elements.

First, I know that 25% of a 20-track album is 5 songs. To make things easier, I'll call the songs A, B, C, D, and E. These will be my elements. I also know from reading the problem that song A must always be the first song in the order. So really there are only 4 elements. The beginning elements are the 4 songs:

song B (element #1)   song C (element #2)   song D (element #3)   song E (element #4)

Step 2: Select the first element and list every possible combination you can create with it.

I'll start with the possible combinations when the first song is A and the second song is ***B****:*

A,**B**,C,D,E

A,**B**,C,E,D

A,**B**,D,C,E

A,**B**,D,E,C

A,**B**,E,C,D

A,**B**,E,D,C

(That's 6 possible combinations so far.)

Step 3: Select a second element and list every possible combination you can create with it.

Next, I'll start with the possible combinations when the first song is A and the second song is ***C****:*

A,B,C,D,E

A,B,C,E,D

A,B,D,C,E

A,B,D,E,C

A,B,E,C,D

A,B,E,D,C

(That's 6 possible combinations so far.)

A,**C**,B,D,E

A,**C**,B,E,D

A,**C**,D,B,C

A,**C**,D,C,B

A,**C**,E,B,C

A,**C**,E,C,B

(That's 6 more possible combinations, for a total of 12 so far.)

Answer Guide Step 4 continued on next page

Step 4: Select a third element and list every possible combination you can create with it. Keep a lookout for combinations you have already made.

Next, I'll start with the possible combinations when the first song is A and the second song is **D**:

A,B,C,D,E	A,C,B,D,E	A,**D**,B,C,E
A,B,C,E,D	A,C,B,E,D	A,**D**,B,E,C
A,B,D,C,E	A,C,D,B,C	A,**D**,C,B,E
A,B,D,E,C	A,C,D,C,B	A,**D**,C,E,B
A,B,E,C,D	A,C,E,B,C	A,**D**,E,B,C
A,B,E,D,C	A,C,E,C,B	A,**D**,E,C,B
(That's 6 possible combinations so far.)	(That's 6 more possible combinations, for a total of 12 so far.)	(That's 6 more possible combinations, for a total of 18 so far.)

Step 5: If there are more than 3 elements, repeat the process until all elements have been used.

Next, I'll start with the possible combinations when the first song is A and the second song is **E**:

A,B,C,D,E	A,C,B,D,E	A,D,B,C,E	A,**E**,B,C,D
A,B,C,E,D	A,C,B,E,D	A,D,B,E,C	A,**E**,B,D,C
A,B,D,C,E	A,C,D,B,C	A,D,C,B,E	A,**E**,C,B,D
A,B,D,E,C	A,C,D,C,B	A,D,C,E,B	A,**E**,C,D,B
A,B,E,C,D	A,C,E,B,C	A,D,E,B,C	A,**E**,D,B,C
A,B,E,D,C	A,C,E,C,B	A,D,E,C,B	A,**E**,D,C,B
(That's 6 possible combinations so far.)	(That's 6 more possible combinations, for a total of 12 so far.)	(That's 6 more possible combinations, for a total of 18 so far.)	(That's 6 more possible combinations, for a total of 24 different combinations.)

Step 6: Make sure you have eliminated duplicates before submitting a final answer.

There are no duplicates, so there are a total of 24 different combinations.

Step 7: How do you know your answer is reasonable?

(Answers will vary.) There are 4 different songs that can be put in order after song A. After the second song is determined, there are 3 different songs that can be put in order. After the third song, there are 2 different songs that can be put in order. After the fourth songs, there is only 1 song left to be put in order. Since 4 x 3 x 2 x 1 = 24, I know my answer is reasonable.

HIP-HOP MATH STUDENT CHALLENGE

The Case of the Missing Money

PLATINUM PROBLEM

Scenario: Kanye West invites DJ TripppleBeatz out to California to be his guest at the Grammy awards and tells DJ TripppleBeatz that he can bring two friends with him to the show. The three men decide to share a suite at a luxury hotel on Sunset Boulevard that costs $300 for the night. They pay in cash and split the bill equally.

After they check in and go up to their room, the hotel manager discovers that these men are friends with Kanye West and reduces the cost of the room to $250 for the night because he wants the musician's future business at his establishment. The manager gives the bellboy $50 in cash and tells his bellboy to return the extra money to the three men.

The bellboy, however, decides to cheat DJ TripppleBeatz and his friends and pockets $20.00, giving each of the men only $10 apiece.

On the way down the elevator to return to the front desk, the bellboy realizes that each man has now paid $90.00 to stay in the room and 3 x $90.00 = $270.00. Plus, the bellboy knows he has pocketed $20.00. After doing the math in his head, $270.00 + $20.00 = $290.00, the bellboy asks himself, "Where is the missing $10.00?"

What is the answer to the bellboy's question?

HIP-HOP MATH ANSWER GUIDE

The Case of the Missing Money

PLATINUM PROBLEM

Scenario: Kanye West invites DJ TrippppleBeatz out to California to be his guest at the Grammy awards and tells DJ TrippppleBeatz that he can bring two friends with him to the show. The three men decide to share a suite at a luxury hotel on Sunset Boulevard that costs $300 for the night. They pay in cash and split the bill equally.

After they check in and go up to their room, the hotel manager discovers that these men are friends with Kanye West and reduces the cost of the room to $250 for the night because he wants the musician's future business at his establishment. The manager gives the bellboy $50 in cash and tells his bellboy to return the extra money to the three men.

The bellboy, however, decides to cheat DJ TrippppleBeatz and his friends and pockets $20.00, giving each of the men only $10 apiece.

On the way down the elevator to return to the front desk, the bellboy realizes that each man has now paid $90.00 to stay in the room and 3 x $90.00 = $270.00. Plus, the bellboy knows he has pocketed $20.00. After doing the math in his head, $270.00 + $20.00 = $290.00, the bellboy asks himself, "Where is the missing $10.00?"

What is the answer to the bellboy's question?

(Answers will vary.)

Be careful about accepting the bellboy's logic! The flaw is in his phrase "plus the bellboy knows he pocketed $20.00." The bellboy should have realized that each man has paid $90.00 to stay in the room and 3 x $90.00 = $270.00. Plus, the bellboy knows he has pocketed $20.00. Simple subtraction shows that $270.00 – $20.00 = $250.00, the exact amount that the hotel has charged!

Written out as an organized list:

Action	Hotel	Bellboy	Men	Total Money
Men have not yet paid	$0	$0	$300	$300
Men pay hotel	$300	$0	$0	still $300
Manager gives $50 to bellboy	$250	$50	$0	still $300
Bellboy gives $10 to each man ($30 total)	$250	$20	$30	still $300!

TWELVE TOP TERMS: KEY ACADEMIC VOCABULARY

Arrange
To put things in a certain order.

Combination
A sequence of numbers or letters.

Duplicate
To make or be an exact copy of something.

Element
An essential or important part of something.

Evaluate
To find a numerical expression.

Exclusively
To be the only one.

Forthcoming
Planned for, or about to happen in the near future.

Maximum
The greatest or highest amount possible or attained.

Narrow
To become or made more limited or restricted.

Options
Choices or possibilities to choose from.

Original
The first or earliest of something; sometimes used as a model for imitation.

Reduce
To make smaller or less in amount, degree, or size.